Data Quality Management in Wireless Sensor Networks

Kewei Sha

Data Quality Management in Wireless Sensor Networks

A Consistency Model Based Approach

VDM Verlag Dr. Müller

Impressum/Imprint (nur für Deutschland/ only for Germany)
Bibliografische Information der Deutschen Nationalbibliothek: Die Deutsche Nationalbibliothek verzeichnet diese Publikation in der Deutschen Nationalbibliografie; detaillierte bibliografische Daten sind im Internet über http://dnb.d-nb.de abrufbar.
 Alle in diesem Buch genannten Marken und Produktnamen unterliegen warenzeichen-, marken- oder patentrechtlichem Schutz bzw. sind Warenzeichen oder eingetragene Warenzeichen der jeweiligen Inhaber. Die Wiedergabe von Marken, Produktnamen, Gebrauchsnamen, Handelsnamen, Warenbezeichnungen u.s.w. in diesem Werk berechtigt auch ohne besondere Kennzeichnung nicht zu der Annahme, dass solche Namen im Sinne der Warenzeichen- und Markenschutzgesetzgebung als frei zu betrachten wären und daher von jedermann benutzt werden dürften.

Coverbild: www.purestockx.com

Verlag: VDM Verlag Dr. Müller Aktiengesellschaft & Co. KG
Dudweiler Landstr. 99, 66123 Saarbrücken, Deutschland
Telefon +49 681 9100-698, Telefax +49 681 9100-988, Email: info@vdm-verlag.de
Zugl.: Detroit, Wayne State University, Diss., 2008

Herstellung in Deutschland:
Schaltungsdienst Lange o.H.G., Berlin
Books on Demand GmbH, Norderstedt
Reha GmbH, Saarbrücken
Amazon Distribution GmbH, Leipzig
ISBN: 978-3-639-17793-0

Imprint (only for USA, GB)
Bibliographic information published by the Deutsche Nationalbibliothek: The Deutsche Nationalbibliothek lists this publication in the Deutsche Nationalbibliografie; detailed bibliographic data are available in the Internet at http://dnb.d-nb.de .
Any brand names and product names mentioned in this book are subject to trademark, brand or patent protection and are trademarks or registered trademarks of their respective holders. The use of brand names, product names, common names, trade names, product descriptions etc. even without a particular marking in this works is in no way to be construed to mean that such names may be regarded as unrestricted in respect of trademark and brand protection legislation and could thus be used by anyone.

Cover image: www.purestockx.com

Publisher:
VDM Verlag Dr. Müller Aktiengesellschaft & Co. KG
Dudweiler Landstr. 99, 66123 Saarbrücken, Germany
Phone +49 681 9100-698, Fax +49 681 9100-988, Email: info@vdm-publishing.com

ACKNOWLEDGEMENTS

During my pleasant work in the Mobile and Internet Systems Laboratory at Wayne State University, I have got great help from many people. I would like to thank them at this opportunity. First, I would like to thank my advisor, Dr. Weisong Shi, for his invaluable advise, support, and discussion in my research. His vision, intelligence and research experience are an inspiration and a model for me. I thank Dr. Loren Schwiebert, Dr. Monica Brockmeyer and Dr. Cheng-zhong Xu to be on my dissertation committee and their treasure comments and suggestions. I also thank Yong Xi, Guoxing Zhan and Safwan Al-Omari, Sivakumar Sellamuthu, Junzhao Du for their great support. I am also thankful to my friends, , Zhengqiang Liang, Hanping Lufei, Zhaoming Zhu, Hui Liu, Yonggen Mao, Sharun Santhosh, Brandon Szeliga, Chenjia Wang and Tung Nguyen for their friendship and support. Finally, I would like to thank my parents and my wife, Jie, for their great support.

TABLE OF CONTENTS

Chapter   **Page**

LIST OF TABLES

LIST OF FIGURES

CHAPTER 1

INTRODUCTION

As new fabrication and integration technologies reduce the cost and size of micro-sensors and wireless sensors, we will witness another revolution that facilitates the observation and control of our physical world [Akyildiz et al., 2002, Estrin et al., 2002, Estrin et al., 1999, Estrin et al., 2003, Pottie and Kaiser, 2000] just as networking technologies have changed the way individuals and organizations exchange information. Micro sensors such as Motes from Intel and Crossbow [crossbow,] have been developed to make wireless sensor network applications possible; TinyOS [Hill et al., 2000, Levis et al., 2003] has been designed to provide adequate system support to facilitate sensor node programming; and finally, several efficient protocols have been proposed to make the sensor system workable. Several applications, such as habitat monitoring [Szewczyk et al., 2004a, Szewczyk et al., 2004b], ZebraNet [Sadler et al., 2004], Counter-sniper system [Simon et al., 2004], environment sampling [Batalin et al., 2004], target tracking [Shrivastava et al., 2006] and structure monitoring [Xu et al., 2004], have been launched, showing the promising future of wide range of applications of wireless sensor networks (WSNs). Recently, several novel architectures have been funded by NSF, such as SNA [SNA,] at UC Berkeley, Tenet [Gnawali et al., 2006] at UCLA/USC, COMPASS [COMPASS,] at Rice, and Wavescope [Wavescope,] at MIT.

Sensing and communication technologies together broaden the way we observe and serve the world, thus in recent several years, more and more physical systems start to integrate those latest technologies into their systems so that they can take advantage of power of sensing and communication [Atkins, 2006, Campbell et al., 2006, Douglas et al., 2006, Garrett et al., 2006, Krause et al., 2008, Srivastava et al., 2006], which not only brings new opportunities to extend the application of traditional wireless sensing systems but also brings new challenges in designing of the extended wireless sensing systems. For example, after sensors and communication components are installed on vehicles, vehicles can form a extremely large-scale high-mobility ad hoc sensing system, Vehicular Networks. Based on vehicular networks, real-time traffic information and interested environmental information can be collected, and various services such as safety services, dynamic routing services can be deployed. With more and more traditional biomedical sensors equipped with wireless communication components, those sensors together with smart phones and the existing cellular or WLAN networks form a healthcare personal area sensing network, which can collect biophysical data from the patient and perform the function of long-term body condition monitoring.

Among all aforementioned systems, we find that different system architectures as well as variant system protocols are requires to cater the requirements of different wireless sensing system applications. As a result, an application-specific approach will be explored in those system design. Whereas, we also envision that, among all aforementioned systems, the success of those wireless sensing system applications is nonetheless determined by whether those wireless sensing systems can provide a *high quality stream of data* over a long period. The inherent feature of deployment of those wireless sensing systems in a malicious or uncontrolled environment, however, imposes challenges to the underlying systems. These challenges are further complicated by the fact that sensor systems are usually seriously energy constrained. However, this problem is largely neglected in previous research, which usually focuses on devising techniques to save the sensor node energy and thus extend the lifetime of the whole wireless sensing systems. However, with more and more deployments of real sensing systems, in which the main function is to collect interesting data and to share with peers, data quality management has been becoming a more important issue in the design of sensor systems. In this dissertation, we envision that data quality management should be integrated in an energy-efficient all sensing system design.

We argue that *the quality of data should reflect the timeliness and accuracy of collected data that are presented to interested recipients who make the final decision based on these data.* Therefore, new protocols are needed to collect both fresh enough and accurate enough data in an efficient way, and the task of deceptive data detection and filtering plays a vital role in the success of data collection. We believe that security using traditional cryptography mechanisms, such as encryption for confidentiality, hashing digest for message integrity, are definitely important and necessary, but not enough to detect deceptive data resulting from data collection and data transmission through multi-hop wireless links. In this dissertation, we undertake a novel approach that detects and filters deceptive data through considering the consistency requirements of data, and study the relationship between the quality of data and the multi-hop communication and energy-efficient design of wireless sensing systems.

To achieve the goal of collecting high quality data in an efficient way, we first analyze a set of real-world sensing data from an environmental monitoring application, then we formally define a new metric, *data consistency*, against which the quality of data is evaluated and the deceptive data is detected and filtered. Intuitively most people think that the high requirements of data quality, the more energy will be consumed. However, based on our observation from sensing data analysis, we find that in most cases the energy could be saved if we consider data consistency and data dynamics together, which inspired us to attack the problem from the prospective of data consistency and data dynamics, and exploit the data consistency in the system protocol design. Second, we propose a set of APIs to manage the consistency and devise an adaptive protocol to integrate data dynamics with data consistency. Moreover, we design a general framework to detect and filter deceptive data so that the quality of the collected data can be largely improved. Finally, energy efficiency is still one of major design goals in our design. Thus, we argue that a good metric is necessary to systematically evaluate the energy efficiency performance of the proposed protocols.

All these models and protocols are integrated in a framework named *Orchis*, which basically has six components, including an analysis to the characteristics of the sensing data from an environmental application, a set of data consistency models customized to wireless sensing systems, a set of APIs to management the quality of collected data, an adaptive protocol for data sampling, a framework to detect and filter deceptive data, and a formal model for the lifetime of the wireless sensing system to evaluate the energy efficiency performance of the protocols. Specifically, we have investigated the following six research problems in this dissertation:

1. *Model lifetime for wireless sensor networks.* Although in the extended sensing systems, the energy constraint is not as severe as that in traditional sensing systems, energy efficiency is still one of the major goals in sensing system design; We propose a novel model to formally define the lifetime of a wireless sensor network based on energy by considering the relationship between individual sensors and the whole sensor network, the importance of different sensors based on their positions, the link quality in transmission, and the connectivity and coverage of the sensor network. Based on our model, we have compared two types of query protocols, the direct query protocol and the indirect query protocol, in terms of mathematical analysis. Then, a comprehensive simulation is done to validate the correctness of the mathematical analysis built on our model. The simulation results shows the correctness of our model.

2. *Analyze the characteristics of a set of sensing data collected in an environmental monitoring application* it is crucial to carefully analyze the collected sensing data, which not only helps us understand the features of monitored field, but also unveil any limitations and opportunities that should be considered in future sensor system design. In this dissertation, we take an initial step and analyze one-month sensing data collected from a real-world water system surveillance application, focusing on the data similarity, data abnormality and failure patterns. Based on our analysis, we find that, (1) Information similarity, including pattern similarity and numerical similarity, is very common, which provides a good opportunity to trade off energy efficiency and data quality; (2) Spatial and multi-modality correlation analysis provide a way to evaluate

data integrity and to detect conflicting data that usually indicates appearances of sensor malfunction or interesting events.

3. *Model the data consistency in wireless sensor networks.* Like those in traditional distributed systems, consistency models in wireless sensing systems are proposed for evaluating the quality of the collected data. Based on our knowledge, we are the first to raise the data quality problem in wireless sensing systems. In this dissertation, we propose a novel metric, *data consistency,* to evaluate the data quality. Our consistency models consider three perspectives of consistency: *temporal, numerical,* and *frequency,* covering both individual data and data streams. Furthermore, we also defined spacial and multi-modality consistency for sensing data.

4. *Develop a set of APIs to manage data consistency and handle the deceptive data.* A set of APIs are designed to distribute data consistency requirements to the monitoring area when the sensor network is deployed. Later on, the consistency requirement should be updated according to the observed data consistency from recent collected data. These APIs are essential and enable application scientists to disseminate consistent requirements, check consistency status, manage consistency requirements, and detect and filter deceptive data. Also, these APIs provide interfaces for lower layer data collection protocols to efficiently transfer the data to the sink.

5. *Devise an adaptive protocol to detect deceptive data and improve the quality of collected data and take advantage of data consistency by considering data dynamics.* We observe that data consistency and energy efficiency are closely related to data dynamics. Thus, models for data dynamics are designed. The protocol that automatically adapts the data sampling rate according to the data dynamics in the data field and data consistency requirements from the application is proposed to improve the quality of collected data and to save energy. As a result, energy can be saved when the data dynamics are low or when data consistency requirements are low. Furthermore, the

zoom-in feature of the adaptive protocol helps us not only to detect interests data changes which usually means abnormal data or interested event, but also to detect deceptive data and improve the quality of sensed data significantly.

6. *Design a deceptive data detection protocol to support data consistency and filtering of deceptive data.* The quality of the collected data are mainly affected by the deceptive data, which usually comes from two sources, wrong readings resulted from inaccurate sensing components and unreliable wireless communication, and false data inserted by malicious attackers. In this dissertation, we propose a general framework to detect the deceptive data from the view point of data itself. Basically, we try to filter two types of deceptive data, redundant data and false data. In our framework, those two types of deceptive data are treated differently. Quality-assured aggregation and compression is used to detect and filter redundant data, while role-differentiated cooperative deceptive data detection and filtering and self-learning model-based deceptive data detection and filtering are utilized to filter false data. Finally, when both types of deceptive data are checked and recognized after the data are delivered to a central server, a spatial-temporal data consistency checking can be performed to further detect and filter the remaining deceptive data.

The proposed framework is close-loop feedback control to manage data quality of the collected data, thus it is a general framework that can be used in all kinds of sensing systems and data quality can be bounded by consistency models or application requirements. With Orchis framework, we can expect that high quality data can be collected in an efficient way. The the efficiency and effectiveness of proposed protocols and models in the framework are validated by both simulation and prototypes, which can be found in the different chapters in this dissertation.

The rest of the dissertation is organized as follows. Background about wireless sensor networks and some related work are listed in Chapter 2. Related work is listed in Chapter 3.

Chapter 4 describes the Orchis framework and an overview of all the components in Orchis. In Chapter 6, we analyze the characteristics of the sensing data collected from a real-life environmental monitoring application. Model for lifetime of wireless sensor networks is formally defined in Chapter 5 and a set of consistency models is formally defined in Chapter 7. Chapter 8, an adaptive, lazy, energy-efficiency protocol is proposed, followed by the design of a framework to detect and filter deceptive data in Chapter 9. Finally, conclusion is drawn and future work is listed in Chapter 10.

CHAPTER 2

OVERVIEW OF SENSING SYSTEMS

Integrating the latest technology of sensing, wireless communication, embedded systems, distributed systems, and networking, sensing systems become a very hot research area for both scientists and engineers. A lot of applications have been developed based on sensing systems. Originally, sensing systems are mostly applied in environmental applications such as [Estrin et al., 2003, Martinez et al., 2004]and [Batalin et al., 2004]. When sensors are equipped at vehicles and other devices, the traditional sensing systems have been extended to more complicated sensing systems, including Vehicular Networks and healthcare personal area networks. From the evolution of the sensing systems, we find several trends in the progress, as listed as follows.

First, one significant difference between the traditional sensing systems and the extended sensing systems is the targeting applications. The traditional sensing systems are mostly applied in scientific applications such as environmental monitoring, but the extended sensing systems are more targeting to personal and social applications, and also they are targeting to more commercial applications. As a result, the data flow is one direction for collecting data in traditional sensing systems, while they are two directions for collecting and distributing data in the extended sensing systems because of the requirements on data sharing. In addition, because of the involve of persons, security and privacy play a bigger concern in the extended sensing system design. Second, the traditional sensing system are usually applied for single dedicated application, while the extended sensing system will mostly be applied by an integration of multiple applications. For example, an environmental monitoring sensing system will focus on collecting data from monitoring field, while a vehicular network can be used both in collecting real-time road information and in deploying a lot of location-based services. Third, sensors in the traditional sensing systems are mostly static, while in the extended sensing systems such as vehicular networks and personal area networks, most sensors have high mobility due to the movement of the vehicles and persons. The mobility of those sensors introduces both challenges and opportunities. Forth, in most environmental applications, sensors are distributed in an uncontrolled area and energy efficiency is a big problem. In the extended sensing systems, sensors are more likely to be deployed in a controlled space, and some sensors have chance to be recharged, so although energy efficiency is still a big design concern in extended sensing systems, the limitation on power supply is not as severe as that in traditional sensing systems. Last but not least, in traditional sensing systems, most sensors are homogeneous, while they are more heterogeneous in the extended sensing systems. Furthermore, the extended sensing systems will have more closer entanglement with the current existing networks such as Internet and Cellular Networks, thus, it more heterogeneous communication protocols will be involved.

Because of the different focuses and requirements in various sensing applications and various sensing systems, the design of the sensing system usually follows an application specific approach, i.e., the system design is largely adapted to the requirements of the applications; however, we argue that there are still some common concerns in sensing system design. For example, energy efficiency is one of the most important concerns in sensing system design, because most sensors are battery powered. Although energy efficiency problem has been relieved somehow in the extended sensing systems such as vehicular networks, it is still a concern even in the design of those systems. Security and privacy are another big concerns in the system design, especially in the extended sensing systems such as healthcare personal-area networks, in which privacy may be the key to the acceptance of those sensing systems. Furthermore, we argue that the most important concern in all sensing system design should be data quality management, because the major function of sensing system is to collect interested high quality data. In this dissertation, we try to integrate the data quality management in an energy efficient sensing systems design.

Having known that there are different sensing systems, such as traditional sensing systems, and the extended sensing systems. Here we give an overview of those existing sensing systems, including traditional sensing systems, vehicular networks, and healthcare personal-area sensing networks. In the following sections, we will introduce the architecture and the characteristics of them one by one.

2.1 Traditional Wireless Sensing Systems Overview

Traditional wireless sensing systems usually have two ways to organize the sensors in the wireless sensor network. One is a flat structured wireless sensor network. The other is a hierarchical structured wireless sensor network. The overview of these two types of networks are listed in next two subsections.

Traditional wireless sensor networks can be applied in various scenarios. We list several typical applications of the traditional sensing systems. For instance, they are used a lot in passive monitoring applications such as traffic monitoring [Skordylis et al., 2006], battle field monitoring [Bokareva et al., 2006, Sha and Shi, 2006b] and environment monitoring [Batalin et al., 2004, Szewczyk et al., 2004b]. They are also used a lot in event-driven applications like enemy detection, fire detection, and object tracking. Furthermore, they are very useful in active querying applications, e.g., you can check the condition of the object you are monitoring by sending it some queries. These different applications may have various specific application requirements to the sensor system design. Thus, traditional wireless sensor networks can be heterogenous systems because of specific application requirements, i.e., in different applications different types of devices may be used as sensors. For example, the motes from CrossBow and Intel can be used as sensors; the RFID readers can act as sensors and collect information from RFID; moreover, the device like web-camera can be used as sensors to monitor the activity of the area. In addition, sensors can be equipped with GPS devices to get accurate location. The two major structure of the sensing systems are listed as follows.

2.1.1 Flat Structured Wireless Sensor Networks

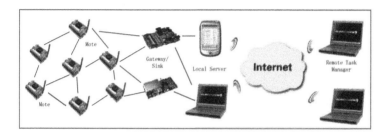

Figure 2.1: Flat structured wireless sensor networks.

A flat structured wireless sensor network usually consists of a set of sensors, a gateway or sink and some other powerful machines as shown in Figure 2.1. Each sensor has the similar capability of computation, communication, and sensing. With the ability of wireless communication, sensors can self-organized into a network that usually has flexible structure. Each sensor in the sensor network may perform as a router so that it forwards message for others. In the data collection process in sensor networks, the data is first sampled at each sensor, then it is transferred to a gateway or a sink, which is usually a more powerful device, in a multi-hop way. The data are further delivered to some local servers such as local servers like desktops and Laptops, which have more powerful computing capability and larger storage, or remote machines via Internet, which locate far away from the sensor network and act as remote task management.

2.1.2 Hierarchical Structured Wireless Sensor Networks

Although the principle of hierarchical tiered structured wireless sensor network is proposed recently in [Gnawali et al., 2006], it has been used in several previous real wireless sensor network deployments, such as [Arora et al., 2005, Szewczyk et al., 2004b]. The basic idea of the hierarchical structured wireless sensor network is show in Figure 2.2.

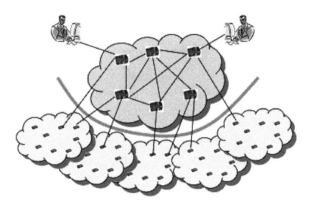

Figure 2.2: Hierarchical structured wireless sensor networks.

In the figure, we can see that sensors are grouped into two tiers, master tier denoted by the bigger motes and mote tier denoted by the smaller motes. The sensor nodes at the mote layer are regular cheap sensors such as MICA2, TelsB motes, which are highly resource constrained. The sensor nodes in the master layer are usually much powerful sensors such as Stargates, which have less battery constraint, more memory, faster processor, more reliable communication. So that most computing intensive tasks, such as data aggregation and mining, are assigned to these powerful motes. This kind of wireless sensor networks has the advantage of more reliable and more manageable.

2.1.3   Characteristics of Traditional Wireless Sensor Networks

Although sensor networks are a special type of traditional distributed systems and mobile ad-hoc networks, there are several major differences between wireless sensor networks and traditional distributed systems. These differences are listed as following.

- The number of sensor nodes is much more than that in the other systems, i.e., it is common to have thousands of sensors in one sensor network.

- Sensor nodes are usually deployed from an aircraft in a rapid, ad hoc manner, so that the deployment and the topology of the sensors are out of control at most time.

- The sensor nodes are usually densely distributed but they are more likely to fail due to the physical failure or out of power supply.

- Sensors are resource constrained devices which have limited bandwidth, computation capability, and small memory, which probably will disappear gradually with the development of new fabrication technologies. The most significant difference between the wireless sensor network and the traditional distributed system is the limited energy supply of sensors, which is believed to be a long term problem.

- It is difficult to replace failed sensors in the sensor network because sensors may be deployed to an area where human beings cannot access, such as in hazardous area and waste containment cover clay soil.

- Sensor networks are application specific systems, which are applied in various scenarios so that they may have different design goals and requirements.

- Data aggregation is a common strategy used in wireless sensor networks to reduce the total amount of data traffic. Thus, aggregation is much more in wireless sensor networks than in traditional distributed systems.

- Communication in the wireless sensor network is not as reliable as traditional networks. So services such relay service as described in [Du et al., 2005] are needed.

These differences make the previous routing protocols and approaches in the wireless ad hoc networks or distributed systems not sufficient to be used in wireless sensor networks. How to optimize the sensor network operations has become a very hot research topic.

2.1.4 Usage Pattern of Wireless Sensor Networks

The main function of a sensor network is to monitor and gather data from the sensor field. Generally, there are three usage patterns of a sensor network to collect data: active querying (pull), passive monitoring (push), and a combination of these two. In the case of active querying, each time data is needed, the sink generates a query message to ask for data from the sensor network. Sensors with corresponding data generate a reply message and route it back to the sink or gateway. For example, if the sink wants the current temperature of an area that is monitored by several sensors, it sends a query to these sensors asking for the temperature. In the passive monitoring mode, however, the sensors are gathering readings at all times and periodically report these readings to the sink. For example, a sensor used to monitor the humidity of some area and required to report the humidity once an hour, or report interesting events such as when the humidity exceeds a pre-defined maximum. This is also known as event driven sensor usage. In addition, a combination of both push and pull is used in some scenarios. When the sink sends queries to ask for specific sensor readings, these queries are active for a period of time during which sensors having relevant data for that query report these readings to sink. For example, sensors collect the wind speed in some area. Under normal weather conditions, the sink queries the wireless sensor network for the wind speed as needed. If the weather becomes violent, the sink may send out a query asking the sensors to report the wind speed every 30 minutes for two days. The difference between active querying and passive monitoring lies in the time interval during which the query is valid. In active querying, the query is satisfied after the reply is sent.

On the other hand, the query will last for a long time in passive monitoring during which time reply messages are sent repeatedly. With hybrid monitoring, the query duration varies; sometimes the query lasts a long time and other times it quickly becomes invalid.

2.1.5 Communication Models in Wireless Sensor Networks

The main usage of wireless sensor network is to collet the data from the monitoring field. Usually, there are four communication models used in wireless sensor network to collect data. We can classify the communication models in sensor networks into four types: Unicast, Area Multicast, Area-Anycast, and Broadcast. These four communication models are abstracted to fit the characteristics of the data source. The difference among the four communication models lies in the granularity of the area of the data source. In the unicast model, the data source of a query is an individual sensor, so the communication is just point-to-point communication between the sink and one sensor, e.g., the query is delivered from the sink to the sensor and the data is collected in the reverse way. Area multicast is used when application is interested in the data from a certain area, so it routes the query to all sensors in a certain area, and then all the sensors in the area generate a reply message to the sink. Alternately, area-anycast is also interested in readings in a certain area, so it routes the query to a specific area and at least one sensor in this area sends a reply message to the sink. Finally, in the case of broadcast, the query message is routed to every sensor in the network, and all sensors with corresponding data return a reply to the sink. These four communication models can be used in the usage patterns described in previous subsection.

2.1.6 Two Query Protocols

In the wireless sensor network, there are two major query protocols as depicted in [Sha et al., 2006b]. These two query protocols are named direct query denoted as Traditional and indirect query denoted as IQ. In direct query protocol, the query is directly sent from sink to the sensors that may have the corresponding data, and the sensors that do have the corresponding data will send the data back to sink. While this approach has a problem of load imbalance, so IQ is proposed to balance the load. The basic idea of IQ can be described as the following three steps. First, the data sink randomly selects a sensor as the query delegate and forwards the query to the delegate. Second, the delegate gets the query and conducts the query processing on behalf of the data sink, and then aggregates the replies. Third, the delegate sends the aggregated reply back to the data sink. Comparing with the traditional query model, two extra steps, query forwarding and query replying, are added in the IQ protocol. Moreover, for point-to-point communication pattern (as Unicast), the performance of IQ is the same as that of the traditional model by choosing the sink as the delegate, but for point to area multicast (and broadcast), such as direct diffusion and flooding (broadcast) based approaches, it will be very helpful.

2.1.7 Routing Protocols

In traditional sensing systems, routing protocol has been extensively explored. Among them we lists several typical routing algorithms as follows.

Greedy geographic routing. GEAR [Yu et al., 2002] and GPSR [Karp and Kung, 2000] are two greedy geographic routing protocols that are close to our work. Both of them have not consider the global information and the local hole information. Especially, GPSR is a purely greedy geographic routing protocol. Furthermore, the derived planer graph in GPSR is much sparser than the original one, and the traffic concentrate on the perimeter of the sparser planar graph in the perimeter node using GPSR make node on planar graph depleted quickly. Thus, they are not so load balanced and fault tolerant.

Fault tolerant routing. A novel general routing protocol called WEAR [Sha et al., 2006a] is then proposed by taking into consideration four factors that affect the routing policy, namely *the distance to the destination, the energy level of the sensor, the global location information*, and *the local hole information*. Furthermore, to handle holes, which is a large space without active sensors caused by fault sensors, WEAR propose a scalable, hole size oblivious hole identification and maintenance protocol. Gupta and Younis propose a fault-tolerant clustering in [Gupta and Younis, 2003]; Santi and Chessa give a fault-tolerant approach in [Santi and Chessa, 2002]. Both of them try to recover the detected faulty nodes, which is actually infeasible when WSN is deployed to a forbidden place. Another fault tolerant protocol by Datta is posted in [Datta, 2003], but it is specific for a low-mobility and single hop wireless network. On the other hand, in WEAR, sensors try to avoid routing messages to a failed field. Other work such as fault tolerant data dissemination by Khanna et al. [Khanna et al., 2004] uses multi-path to provide the fault tolerant, which has to keep more system states to achieve the goal.

Information exploiting routing. Data-centric routing such as Direct Diffusion [Intanagonwiwat et al., 2000] use interest to build the gradient and find a reinforced path to collect data. RUGGED [Faruque and Helmy, 2004] by Faruque and Helmy direct routing by propagating the events information. However, all of them pervade useful information. On the contrary, WEAR distributes harmful hole information. Similar to WEAR, GEAR tries to learn the hole information. However, the hole information propagation is much faster and more sufficient in WEAR than that in GEAR. Furthermore, GEAR needs to keep a large amount of information for every destination.

2.2 Vehicular Networks

About half of the 43,000 deaths that occur each year on U.S. highways result from vehicles leaving the road or traveling unsafely through intersections. Traffic delays waste more than a 40-hour workweek for peak-time travelers [VII,]. Fortunately, with the development of micro-electronic technologies and wireless communications, it is possible to install an On-Board-Unit (OBU), which integrates the technologies of wireless communications, micro-sensors, embedded systems, and Global Positioning System (GPS), on vehicles. By enabling vehicles to communicate with each other via Inter-Vehicle Communication (IVC) as well as with roadside units via Roadside-to-Vehicle Communication (RVC), vehicular networks can contribute to safer and more efficient roads by providing timely information to drivers

18

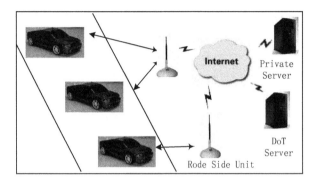

Figure 2.3: A typical authentication scenario.

and concerned authorities. After having more and more vehicles are equipped with the communication devices, the vehicular network is becoming one of the largest ad hoc sensing systems that exist in our daily life.

Vehicular networks can be used to collect traffic and road information from vehicles, and deliver road services including road warning and traffic information to the users in the vehicles. Thus, a great attention has been put into designing and implementing similar systems in the past several years [Bishop, 2000, ITSA and DoT, 2002]. Several typical vehicular network applications are described as follows. First, vehicular networks can be used as traditional sensor networks with higher mobility, thus, vehicles can be used to collect the environmental condition and send them via the inter-vehicle communication and vehicle-to-roadside communication. In vehicular networks, we find that there are mainly two types of data will be collected, each of which has its specific features. One is regular data which includes normal whether conditions, normal road conditions and some information about the vehicle itself such as location, velocity and direction. This type of data usually has low real-time requirements, which is usually used for long-term analysis. The other type of data is events data, which includes sudden brake, heavy traffic, car accident, hard turn, wild whether conditions such as icy and froggy. These kinds of data usually has high real-time requirements, which should also be delivered to give warning to others vehicles on the road. Second, utilizing the collected data, vehicular networks are very helpful to improve the safety in drive. With the inter-vehicle communication, safety warning information can be delivered in a timely way so that the driver can response before severe damage happens. Third, a lot of location-based service can be deployed in vehicular networks. For example, intelligent navigation system can be designed based on the real-time traffic information collected by the vehicular networks. Other location-based road aid service like featured store reminding can also deployed in vehicular networks.

2.2.1 *Architecture of Vehicular Networks*

A vehicular network consists of a set of vehicles on the road, a set of road side unit (RSU), and a number of remote server, as shown in Figure 2.3. Usually, the vehicles will move on the road and collect information such as road conditions, environmental parameters or detected events. Each vehicles acts like a mobile sensor mote with the capability of computation, communication and sensing. RSUs are deployed along the road, acting as fixed sensor motes, which can also capable of computation, communication and sensing. In most cases, the RSUs are more powerful than vehicles in terms of computation and communication. Remote servers are mostly located far away from the road and they are normally much more powerful and have high ability in data processing. The communication happens in several cases, including between RSUs and vehicles, among vehicles and between RSUs and remote servers. Dedicated Short Range Communications (DSRC) is the designated wireless protocol for a vehicular network [DSRC,], which is used in the communication between RSUs and vehicles as well as among vehicles. Internet based communication is exploited to transfer data between RSUs and the remote server as well as among RSUs.

2.2.2 Characters of Vehicular Networks

Compared with other wireless sensor networks, vehicular networks have their special features. First of all, vehicular networks usually include a very large amount of vehicles, but only when these vehicles are driving on road, they are involved in the network. Furthermore, the vehicle is moving very fast on the way, thus the network is highly dynamic. As a result, it lacks permeant relationship among vehicles and RSUs as well as among different vehicles. Second, in such a totally distributed environment, all decisions should be made locally based only on partial information. Moreover, some decisions have to be made in a real-time format. Third, one major application of vehicular networks is to improve the safety, thus safety is a big concern in vehicular networks; then, an emergency event should be reported and confirmed by other vehicles or RSUs on time. Making an accurate real-time decision based on partial information is a big challenge. Also, security will be considered the vehicular network design to prevent malicious attacks. Fourth, because vehicular and drivers are closed entangled, drivers will have big concerns to keep privacy, such as location privacy and service usage pattern privacy. Finally, vehicles are usually installed big and rechargeable batteries, so the energy constraints in the vehicular network is not as high as that in the traditional sensor networks and both communication and computation are also more powerful than the traditional sensor networks.

2.3 Healthcare Personal Area Networks

The medical system has not been able to effectively adapt to the dramatic transformation in public health challenges; from acute to chronic and lifestyle-related illnesses . Although acute illnesses can be treated successfully in an office or hospital, chronic illnesses comprise the bulk of health care needs and require a very different approach. There is overwhelming consensus that the prevention and treatment of cardiovascular disease, diabetes, hypertension, chronic pain, obesity, asthma, HIV, and many other chronic illnesses require substantial patient self-management.

People need to monitor their bodies, reduce physiological arousal when stressed, increase physical activity, and avoid or change harmful environments. Yet, there is a lack of effective and easily deployed tools for self-monitoring, and people often do these tasks poorly, especially people at socioeconomic risk for chronic illness, such as urban minorities. That is, people most at risk for costly chronic illnesses have the least access to self-management tools. Based on current cognitive and behavioral change research, we are convinced that the prevention or treatment of chronic illnesses will be greatly aided by an innovative system that can monitor one's body, behavior, and environment during a person's daily life, and then alert the person to take corrective action when health risks are identified. This goal reflects the view of Microsoft Corporation president Bill Gates, who noted in a recent *Wall Street Journal* article, "What we need is to put people at the very center of the health-care system and put them in control of all of their health information".

Currently, smart phones and personal assistant devices, are widely used in field research to collect information from participants. For example, at random or pre-set times, a person is prompted to respond to questions. Moreover, there are numerous devices available to record physiological responses in real life, such as blood pressure and heart rate. Thus, healthcare personal area networks are developed by utilizing those devices to help people to monitor their body conditions and get some feedback or suggestions from healthcare expertise.

In this dissertation, we introduce SPA, a smart phone assisted chronic illness self-management participatory sensing system. SPA represents a general architecture of a healthcare personal area sensing network. There are three main functions of the system. First, it is used to collect real-time biomedical and environmental data from the participant using sensors, which will be very useful for us to understand the possible causes of the chronic illness. Second, the data analysis and data mining algorithms are used to find time-series rules and the relationship between the biomedical and environmental parameters, which will help health care professionals to design health care plans for specific participants. Third,

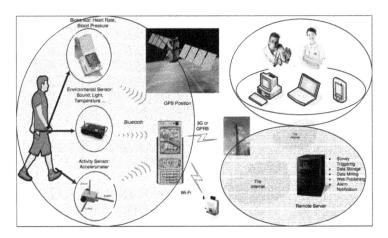

Figure 2.4: The architecture of the SPA system.

the system automatically triggers on-line surveys and sends out alarm notifications, largely reducing the involvement of the health care professionals, which not only saves medical cost but also protects the participants as early as possible. Next, we give a detailed description of SPA system design.

2.3.1 The Architecture of the SPA System

The SPA system consists of three major parts as shown in Figure 2.4, including a body area sensor network to collect biomedical and environmental data, a remote server to store and analyze data, and a group of health care professionals to check records and give health care suggestions.

The body area sensor network includes a smart phone, a set of biosensors and a set of environmental sensors. The smart phone works as a base station for the body area sensor network, which can not only receive and temporarily store sensed data but also work as a router to communicate with the remote server. Moreover, the smart phone is also equipped with a GPS sensor, which can provide location information of the participant and is helpful to context-aware data analysis. A set of biosensors, such as pulse oximeter, blood pressure meter and actigraph, are attached to participant, periodically sampling the heart rate, blood pressure and movement respectively. In addition, environmental sensors are used to sample sound, temperature, humidity, and light. The communication within the body area sensor network is via Bluetooth. A TDMA schedule is devised to collect the body area sensor data. In addition to the biomedical, environmental and location data, the subjective state of the participant is also reported via random or periodical survey questions. Eventually, all sensed data and filled surveys are sent to a remote server by the smart phone.

A remote server is set up to store all collected sensor data. After these data are collected, they will be stored in a formatted data warehouse. Time series data mining algorithms are deployed at the server to discover the time-series patterns and ruled in all collected data. Multi-modality data mining algorithms is utilized to mine the correlation between the biomedical, environmental, and location data. Furthermore, multi-modality data mining algorithms will also detect the conflicts among those data and identify unusual data such as dramatic changes in the sensor readings, which health care professionals should examine and response. All above mentioned algorithms integrate the domain knowledge from health care professionals. Survey data will also be used as a calibration to measure the collected data. Follow-up investigation is necessary whenever there is a mismatch between the survey data and the sensed data. The remote server also takes care of triggering on-line surveys as well as alarm notification to the participant, based on the health care professional's suggestion or detection algorithms. In addition, it presents health records to corresponding legislated health care professionals by using access control mechanisms.

The responsibility of the health care professionals is two-fold. First, they design the context-aware questionnaires and help to design the data mining algorithms using their domain knowledge. Second, the health care professionals need to check the health record from the server periodically, especially for those data which are marked as "conflict" and/or "unusual data". Follow-up suggestions are expected after the health care professionals examine the health record. In our design, trying to reduce the involvement of the health care professionals as much as possible, most surveys and alert notifications, after they are designed, will be automatically generated by the server, based on the predetermined rules and the value of the currently received data. Only when it is in urgent cases will health care professionals be involved to provide support to the system.

The above mentioned three major parts in the SPA system are connected by using a variety of communication protocols. Within the body area network, Bluetooth is adopted to connect the sensors and the smart phone. A TDMA based schedule is designed for the smart phone to collect data from the sensors. If the smart phone is not available during the scheduled time, the sensors will store the data locally and temporarily. Thus, loose synchronization is enough between the smart phone and the sensors. Aggregation algorithms are applied when the volume of the data exceed the size of available storage. The communications between the smart phone and the server can be either WLAN or cellular network based on the network availability and energy concerns. The health care professionals usually access the health record via Internet. The data flow from the sensors to the smart phone, and eventually, data arrive at the remote server. The feedbacks and the questionnaires are initialized by the health care professionals, mostly automatically triggered by the system, and eventually delivered to participants.

2.3.2 Characters of the Healthcare Personal Area Networks

Compared with traditional sensing systems and vehicular networks, healthcare personal area sensing networks have several special features. First, the system is more heterogenous than traditional sensing systems. For example, multiple communication radios may be exploited in the communication. Bluetooth, UWB, and Zigbee protocol may be utilized the communicate the body area sensors and the smart phone or PDAs. Wifi or cellular networks may be used to transfer the information from PDAs to a centralized server. A variant of sensing devices can be used to sense personal biophysical conditions, including blood pressure, heart rate, movement and temperature. Second, The data collected by biosensors and location information are sensitive personal information. Usually, participants are not willing to release such data to the public. Thus, privacy is one of the most important design issues in this type of sensing system. Third, the sensing parameters are more context-aware than sensing parameters in traditional sensing systems. From this perspective, more rich data will be collected. Forth, the communication and data collection in healthcare personal area networks will not take the same strategy as those in most traditional sensor networks. Collected data will be sent mostly in one-hop, but not using multi-hop routing protocols.

CHAPTER 3

RELATED WORK

Having introduced the basic structure and features of the wireless sensor network. In the rest of this chapter, we will introduce some relate work to our design. These previous efforts include the energy efficiency design, the definition for the lifetime of wireless sensor networks, data consistency models, adaptive protocol design, data quality management, and MAC protocol respectively.

3.1 Related Definition of Lifetime of Wireless Sensor Network

Energy-efficient routing protocols and optimizations to maximize the lifetime of sensor networks have been widely studied in the literature [Akkaya and Younis, 2003]; however, few of the previous efforts have been done to formally model the lifetime of the sensor network. To this end, our work is the first step towards this direction.

In several previous work, the lifetime of the sensor network is defined as the time for the first node to run out of power such as in [Chang and Tassiulas, 2000, Heinzelman et al., 2000, Kalpakis et al., 2002, Kang and Poovendran, 2003] or a certain percentage of network nodes to run out of power as in [Duarte-Melo and Liu, 2003, Xu et al., 2001a]. We think that these definitions of the lifetime of the sensor network are not satisfactory. The former is too pessimistic since when only one node fails the rest of nodes can still provide the whole sensor network appropriate functionality. While the latter does not consider the different importance of the sensors in the sensor network, as shown in Section 7.2 the failure sensors in different location will have different influence to the whole sensor network.

In the work of [Bhardwaj and Chandrakasan, 2002, Bhardwaj et al., 2001, Blough and Santi, 2002, Mhatre et al., 2004], the lifetime of the sensor network is defined as the time

when the sensor network first losts connectivity or coverage. The rationale of their definition is based on the functionality of the sensor network, which is similar to our definition.
However the way to detect the termination of the sensor network is different. Blough and
Santi [Blough and Santi, 2002] define it by checking the connectivity of a graph; Mhatre
et al. use a connectivity and coverage model to describe it; while we define it as the time
when the remaining lifetime of the whole sensor network starts to keep constant as losing
connectivity or the sensor network loses coverage.

Xue and Ganz study the lifetime of a large scale sensor network in [Xue and Ganz,
2004]. They explore the relationship between the lifetime of a sensor network with the
network density, transmission schemes and maximum transmission range. Their work is
based on a general cluster-based model, and does not consider the importance of different
sensors. They also aim to explore the fundamental limits of network lifetime. Compared
with their work, our model is more general which can be used not only for cluster-based
model. Furthermore, because we take more factors into consideration in our model, our
model is more useful and flexible, in which the lifetime is calculated according to the really
energy consumption.

Bhardwaj *et al.* define upper bounds on the lifetime of the sensor network in [Bhardwaj
and Chandrakasan, 2002, Bhardwaj et al., 2001]. They explore the fundamental limits of
data gathering lifetime that previous strategies strive to increase. One of their motivations
is to calibrate the performance of collaborative strategies and protocols, but they just give
out an upper bound of the lifetime rather than the actually lifetime model for different
strategies. Besides, our model can also guide the design of the low-level protocols.

Another recent work has been done also aim to derive the upper bound of the lifetime of
a sensor network in [Zhang and Hou, 2004]. The authors want to explore the fundamental
limits of sensor network lifetime that all algorithms can possibly achieve. Compared with
their work, our model is aiming to develop models for both sensor system designer and
application scientist, and we focus on calculating the more accurate lifetime of a sensor

network according to different underlayer routing or query protocols. In both our work, we consider the coverage and connectivity of the sensor network.

Duarte-Melo and Liu provide a mathematical formulation to estimate the average lifetime of a sensor network in [Duarte-Melo and Liu, 2003]. Their work aims to estimate the average lifetime of the sensor network rather than to provide a general model that can be used to measure different protocols. Our model can be used to model the lifetime of the sensor network using different communication patterns, which is more general. In their later work, they give a modeling frame for computing lifetime of a sensor network. Their approach is to maximize the functional lifetime of a sensor network and get the value of it based on the solution of a fluid-flow model. While our goal of this dissertation is to provide a general model for lifetime of sensor network. Besides, the calculation of their lifetime still need a lot of calculation. While our model can be easily calculated based on the centralized algorithms.

Similar to [Duarte-Melo and Liu, 2003], several other efforts such as in [Kalpakis et al., 2002, Kang and Poovendran, 2003, Sadagopan et al., 2003, Mhatre et al., 2004, Shah and Rabaey, 2002] have been done to maximize the lifetime of sensor network. Whereas almost all their work take it as an optimization problem and build a linear programming model, then find an algorithm or a protocol to achieve the maximum lifetime, so that these approaches are always closely related to the routing protocols, rather than giving a general model for the lifetime of sensor network. Besides, most of them ignore the load imbalance problem. Even though some of them do notice the problem, they only balance the load at the routing level.

Energy-aware routing [Shah and Rabaey, 2002] is proposed by Shah *et al.* using a set of sub-optimal paths to increase the lifetime of the network. This approach uses one of multiple paths with a certain probability to increase the lifetime of the whole network. Another similar approach is proposed in [Dai and Han, 2003], which constructs a load balance tree in the sensor networks with load balance to different branches. Their work balances the load of each data path so that extend the lifetime of sensor networks. They do not provide a formal model for the lifetime.

3.2 Related Sensing Data Analysis

SenseWeb [SenseWeb,] has provided a venue for people to publish their data, but we have not seen any analysis yet. Our next step will use more data set from SenseWeb. Data aggregation is an important way to reduce the volume of the collected data. A few data aggregation approaches have been proposed. These approaches make use of cluster based structures [Heinzelman et al., 2002] or tree based structures [Ding et al., 2003, Goel and Estrin, 2003, Intanagonwiwat et al., 2002, Luo et al., 2005, Zhang and Cao, 2004]. Considering applications which require some aggregate form of sensed data with precision guarantees, Tang and Xu propose to differentiate the precisions of data collected from different sensor nodes to balance their energy consumption [Tang and Xu, 2006]. Their approach is to partition the precision constraint of data aggregation and to allocate error bounds to individual sensor nodes in a coordinated fashion.

As the tradeoff between data quality and energy consumption has been considered in a few data aggregation protocols, adaptive sampling provides another way to minimize energy consumption while maintaining high accuracy of data. Essentially, n adaptive sampling design, the sensor nodes which send out data and the rate of data transmission are selected adaptively.

Adaptive sampling has been proposed to match sampling rate to the properties of environment, sensor networks and data stream. Jain and Chang propose an adaptive sampling approach called backcasting [Jain and Chang, 2004], which first makes an initial correlation estimate by having a small subset of the wireless sensors communicate their information to a fusion center, then selectively activates additional sensors in order to achieve a target error level. Gedik and Liu proposed a similar way of data collection, selective sampling [Gedik and Liu, 2005]. Their selective sampling algorithm uses a dynamically changing subset of nodes as samplers to predict the values of non-sampler nodes through probabilistic models. This approach assigns nodes with close readings to the same clusters, and prediction model is used to minimize the number of messages used to extract data from the network. Although many approaches have been proposed to reduce energy while maintaining data quality, there exists rare study on the pattern of raw data collected by sensor nodes in the real world. In addition, most studies adopt the way of simulation, To examine how well those approaches fit our real world, and to inspire new approaches, it's necessary to study the raw data carefully under systematic guidelines, whereas, our quality-oriented sensing data analysis gives a chance to take a fresh look at how the data behaves. We made use of reappearance pattern, empirical distribution, trend comparison, multimodality analysis and spatial correlation to present a clear view of the sensing data, and many interesting facts are discovered. These findings, in turn, enhance our understanding of the reality of today's wireless sensor networks. The idea is to combine the data coming from different sensor nodes, eliminating redundancy, minimizing the number of transmissions and thus saving energy. We are the first that formally define a set of consistency models for WSNs. We also design and implement an adaptive, lazy, energy efficient data collection protocol to improve data quality and save energy.

3.3 Related Energy Efficiency Design

Energy efficiency is always one of the major goals in the design of WSN. Energy efficient protocols have been explored for a long time. Previous work expects to achieve the goal of energy efficiency by designing energy efficient query protocols [Sha and Shi, 2006a], routing protocols, such as [Braginsky and Estrin, 2002, Chang and Tassiulas, 2000, Hamdaoui and Ramanathan, 2006, He et al., 2003, Heinzelman et al., 1999, Heinzelman et al., 2000, Intanagonwiwat et al., 2000, Kalpakis et al., 2002, Lindsey et al., 2002, Lindsey et al., 2001, Madden et al., 2002, Manjeshwar and Agrawal, 2001, Manjeshwar and Agrawal, 2002, Sadagopan et al., 2003, Schurgers and Srivastava, 2001, Seada et al., 2004, Sha and Shi, 2004, Sha and Shi, 2005, Shah and Rabaey, 2002, Xu et al., 2001b, Yao and Gehrke, 2002], energy efficient

MAC protocols like [Polastre et al., 2004, Shih et al., 2001, Ye et al., 2002, Woo and Culler, 2001], energy efficient clustering [Younis and Fahmy, 2004], duty cycle management [Dasgupta et al., 2003, Younis et al., 2002, Younis and Fahmy, 2004], sensor network topology management [Gupta et al., 2008], and other energy efficient approaches [Sadagopan and Krishnamachari, 2004]. However, these approaches mainly focus on finding some energy efficient path, designing better turn on/off schedules, forming energy efficient clusters, and so on, but none of them has examined the energy efficiency from the view of the data itself, i.e., to adapt the data sampling rate to the data dynamics and keep lazy when data consistency is maintained. Thus, we are first to design an energy-efficient protocol from the prospective of data consistency.

Aggregation is one of the most common technologies used in wireless sensor networks to save energy. Aggregation structures such as TAG [Dasgupta et al., 2003, Intanagonwiwat et al., 2002, Madden et al., 2002] are designed to aggregate the message. Another work from [Shrivastava et al., 2004] propose new aggregation scheme that significantly extends the class of queries that can be answered using sensor networks. We also use aggregation in our protocol, and we not only take advantage of the previous aggregation techniques but also try to combine several pieces of data together.

Load balanced protocols are designed in different layer of sensor systems to save energy and extend the lifetime of wireless sensor networks. Dai and Han in [Dai and Han, 2003] construct a load balanced tree in the sensor networks to make the load evenly distribute to different branches of the routing tree. Indirectly Query [Sha et al., 2006b] is proposed to balance the load at query layer so that it extends the lifetime of WSN a lot. At the higher level, several researchers have been proposed to balance the load of sensors by rotating their functionality, including coordinators in topology management [Chen et al., 2001], rotation of grid zone headers in GAF routing [Xu et al., 2001a], rotating cluster headers in hierarchical protocols [Heinzelman et al., 2000], and switching among multi-paths during routing [Ganesan et al., 2002].

Energy aware routing is also used to save energy and extend the lifetime of wireless sensor networks. Shah and Rabaey propose an energy aware protocol [Shah and Rabaey, 2002]. They keep using a set of good paths instead of just finding a single optimal path and use different path at different time with some probability depending on the energy metric. Younis *et al.* [Younis et al., 2002] design a energy-aware routing for cluster-based sensor network. In their approach, the gateway in each cluster applies energy-aware metrics to manage the topology adjustment and routing setup, but the cluster based scheme is argued to be energy inefficient. GEAR [Xu et al., 2001a] and WEAR [Sha et al., 2006a] are both used energy aware approach to balance the load to different sensors so that they extend the lifetime of WSN significantly.

Duty cycle management and sensor network topology management are two other approaches to achieve the goal of energy efficiency. In [Dasgupta et al., 2003, Younis et al., 2002, Younis and Fahmy, 2004], the authors try to design an on/off schedule for sensors so that they can save energy by making sensors sleep as much as possible. However, we argue that the design of duty cycle schedule in sensing systems should be integrated with the data quality management by taking data consistency into consideration. Gupta proposes a topology management mechanism in sensor networks to save energy [Gupta et al., 2008]. The basic idea of this approach is to take advantage of the redundant sensors. To be specific, they pick up a set of sensors from the sensor network and make sure that this set of sensors are sufficient to reconstruct data for the entire sensor networks. This approach can save energy but it also introduce load imbalance, and it cannot satisfy some applications that require a certain level of data redundant. In general, our data quality management scheme complements this kind of approaches very well by considering both energy efficiency and data quality.

3.4 Related Data Consistency Models

Data consistency is a classical problem in computer architecture, distributed systems, database, and collaborative systems. Interesting readers please refer to [Peterson and Davie,

2003, Ramakrishnan, 1998, Tanenbaum and van Steen, 2002]. A lot of consistency models, such as strict consistency, causal consistency, weak consistency, eventual consistency, have been proposed in the research of these fields. However, these models are usually not applicable in WSN. For example, these models are working on how to keep consistency for several duplicated data in the a distributed environment. However, in this dissertation, we do not consider the problem of duplicated data sampled from different sensors but the data consistency between the real time in the data field and the received data at the sink. Thus, these models are not applicable in our scenario.

The work form Yu and Vahdat [Yu and Vahdat, 2000] explores the semantic space between traditional strong and optimistic consistency models for replicated services. They propose an important class of applications can tolerate relaxed consistency, but benefit from bounding the maximum rate of inconsistent access in an application-specific manner. Thus, they develop a set of metrics to capture the consistency spectrum. Similar to us, they also consider the application requirements to the data in the data consistency models. However, their models are still for different replicas and so that they are suitable for using in traditional distributed system instead of being applied in the field of wireless sensor networks.

Ramamritham *et al.* propose an idea to maintain the coherency of dynamics data in the dynamics web monitoring application [Shah et al., 2003]. They address the problem of maintaining the coherency of dynamic data items in a network of repositories using filters. In their follow-up work in [Gupta et al., 2005], they model the dynamics of the data items. Based on the dynamics of data, they adapt the data refresh time to deliver query results with the desired coherency. Their work is similar to ours; for example, both of their work and our work want to model the data dynamics and design adaptive protocol to deliver data; however, their work is to collect data from the web, and our work is to collect data in wireless sensor network, which is more resource constraint. Moreover, we have different goal in data operations from theirs so that we use a different protocol and different models, and our model for data consistency is more general than theirs.

Lu *et al.* propose a spatiotemporal query service that allows mobile users to periodically gather information from their surrounding areas through a wireless sensor network in [Chipara et al., 2005, Lu et al., 2005], and their goal is to provide a service to enable mobile users to periodically gather information and meet the spatiotemporal performance constraints. They propose spatial constraints and temporal constraints to the query and collected data, but in their work they propose neither data consistency models, nor adaptive protocols. Furthermore, their work is most useful in a system with high real-time requirements, while our work defines general consistency model which can be applied in different applications with variant data consistency requirements. However, their work compliments to our effort very well, i.e., we can integrate their approach with our data consistency models by using their service in the scheduling of our protocol. Thus, as far as we know, this is the first model to define the data consistency in WSN.

Information assurance is proposed in [Deb et al., 2003]. Their work intends to disseminate different information at different assurance levels to the end-user so that they can deliver critical information with high assurance albeit potentially at a higher cost, while saving energy by delivering less important information at a lower assurance level. They propose a scheme for information delivery at a desired reliability using hop-by-hop broadcast, based on which they can attain given desired end-to-end reliability. We share the similar goals with their work, i.e., we want to guarantee the end-to-end data consistency and treat different data with different priority according to their application specific consistency requirements; however, they neither abstract the application requirements to the different information nor propose consistency model. Moreover, they use different protocol to collect data in an energy efficient way.

3.5 Related Adaptive Design in Wireless Sensor Networks

Adaptive approaches, which usually fit the changes of the environment or the different requirements of the applications, are always attractive in system design. Several adaptive protocols including [Cerpa and Estrin, 2002, He et al., 2002, Heinzelman et al., 1999, Lin and Gerla, 1997, Mainland et al., 2005] are proposed in literature. However, these protocols mostly focus on the cluster formation, communication patten selection, and duty cycle designing. None of them intends to adapt the data sampling rate according to the data dynamics.

A family of adaptive protocols called SPIN are proposed by Heinzelman *et al.* in [Heinzelman et al., 1999] to efficiently disseminate information in wireless sensor networks. They intend to develop a resource aware and resource adaptive sensor networks, e.g., when the sensors' energy is approaching a low-energy threshold, it adapts by reducing its participation in the protocol. However, we argue that making decision only based on the current available resource is not enough to optimize the performance of the application specific sensor networks, the requirements from application should be considered as well.

ASCENT [Cerpa and Estrin, 2002] is proposed to make sensors coordinate to exploit the redundancy provided by high density, so as to extend overall system lifetime. In ASCENT, each node assesses its connectivity and adapts its participation in the multi-hop network topology based on the measured operating region. Their goal is to adapt the sensor to a wide variety of environment dynamics and terrain conditions by managing the duty cycle. Thus, they are interested in different problems from us. As a result, they intend to extend the lifetime of wireless sensor network by topology control instead of considering data consistency.

In [He et al., 2002] He *et al.* propose a novel aggregation scheme that adaptively performs application independent data aggregation in a time sensitive manner. Their work isolates aggregation decisions into a module between the network and the data link layer. The basic idea of their approach is taking advantage of queueing delay and the broadcast nature of wireless communication to concatenate network units into an aggregate. They design a feedback based adaptive component to make on-line decision on how many packets to aggregate and when to invoke such aggregation based on local current network conditions. Compared with our work, the adaption happens after the data is collected, while in our approach the adaption happens during sampling data. So we can integrate their work and our work together and take advantage of both adaption.

A recent paper [Mainland et al., 2005] from Mainland *et al.* uses adaptive approach to allocate the resource for sensor networks. They model the sensor as a self-interested agent and use price to tune the behavior of each sensor. Their goal is to maximize the profit of each sensors under the constraints of energy budget. In their design, they can also adapt the sampling rate to the data dynamics based on the current energy budget. However, they neither propose a formal model for data dynamics nor consider data consistency in their adaptive approach. Thus their approach is not from the view of the data themselves, but from the view of every sensor's profit.

Several techniques on adaptive sampling rate has been proposed in the database field, these techniques share the same goal of our adaptive protocol. Jain and Chang propose an adaptive sampling approach for WSN [Jain and Chang, 2004]. They employ a Kalman-Filter (KF) based estimation technique and the sensor uses the KF estimation error to adapt the sampling rate. Their approach is different from our approach in that it has to store more data. Moreover Kalman-Filter has matrix operation so that it is much more computing intensive. Marbini and Sacks [Marbini and Sacks, 2003] propose a similar approach to adapt the sampling rate as ours, however they do not model the data dynamics and require an internal model, which is usually difficult to find, to compare the sampled data. TinyDB [Madden et al., 2005] also adapts the sampling rate based on current network load conditions, but not based on the data dynamics in the data field. Their work complements our work very well. More work on sampling rate adaption should be done by considering the network load condition and the data feature such as data dynamics and priorities together.

3.6 Related Data management in Wireless Sensor Networks

Data management has been extensively explored in previous researches. Li *et al.* use a feedback-driven model-based-prediction approach to manage sensed data [Li et al., 2006], which shares similar idea with us. Their work makes a tradeoff between the storage cost and the communication cost; however, they do not provide a general consistency model like us to evaluate data quality, nor do they dynamically adapt the sampling rate to improve data quality.

Trappe *et al* propose MIAMI [Trappe et al., 2005], methods and infrastructure for the assurance of measurement information, to handle the attacks on the process of measurement (PoM attack). They propose a framework based on the development of the PoM monitors, which is responsible for preventing corrupted measurement data. Their framework shares the same goal as u; however, they have not formally define formal consistency models. Furthermore, their data quality management is not a feedback-based control. Instead, Deb *et al* design an information assurance protocol to disseminate different information at different assurance levels by hop-to-hop protocols [Deb et al., 2003], which has a similar goal as our priority differentiated data collection protocol.

Quality aware sensing architecture [Lazaridis et al., 2004]is proposed by Lazaridis *et al* as a scalable data management solution for wireless sensor network applications. They also propose some sophistic compression and prediction algorithms [Lazaridis and Mehrotra, 2003] to manage the volume of the delivered data. They consider data quality from the perspective of timeliness and data accuracy, which is similar as our perspective, but they do not give formal models to describe timeliness and data accuracy. Moreover, they do not distinguish individual data and streaming data. In addition, their approach is more suitable to be used in traditional sensing systems, and they do not detect and filter deceptive data in their architecture. Thus, we believe that their approach is not sufficient to manage the quality of the collected data in sensing systems.

Another work [Larkey et al., 2006] has been done to assure the quality of collected sensing data in an environmental application, by detect measurement errors and infer missing readings. The basic idea of their approach is to take advantage of the spatialtemporal coherence of the sensing data in environmental applications, and estimate distributions to model the data, so this is not a general framework to manage data quality in sensing systems. Confidence-based data management for personal area sensor networks [Tatbul et al., 2004] is proposed to achieve collecting high quality data with a certain confidence. The basic idea is to take redundancy and other factors in to consideration in data value estimation. This is not a general framework and it only works for personal area sensing systems.

Filters are also used to manage data by reducing the size of the data stream. Work by Olston *et al.* uses an adaptive filter to reduce the load of continuous query. Their work focuses on the adaptive bound width adjustment to the filter so that their results are helpful to analyze our lazy approach, but they have not modeled the data consistency and considered adapting sample rate. Sharaf *et al.* study the trade off between the energy efficiency and quality of data aggregation in [Sharaf et al., 2004]. They impose a hierarchy of output filters on WSN to reduce the size of the transmitted data. Data prioritization in TinyDB [Madden et al., 2005] chooses the most important samples to deliver according to the user-specified prioritization function, which is not as general as our work on data consistency and data dynamics. Furthermore, none of protocol related work has proposed model for data consistency, but we explore a whole issue related to data consistency and data dynamics, and our system design is a revision based on these models.

3.7   Related MAC protocols in Wireless Sensor Networks

Previous MAC protocols can be classified into two types, contention-based and reservation-based. IEEE 802.11 DCF [IEEE802.11, 1999] is a contention-based protocol, which builds on MACAW [Bharghavan et al., 1994]. The basic idea of these protocols is for a sender to transmit a request-to-send (RTS) that the receiver acknowledges with a clear-to-send (CTS).If the RTS/CTS exchange is successful, the sender is allowed to transmit one or more packets. CRAM [Garces and Garcia-Luna-Aceves, 1997b] and PRMA [Goodman et al., 1989] try to resolve RTS collisions based on auctions or tree-splitting algorithms.

B-MAC [Polastre et al., 2004], S-MAC [Ye et al., 2002] and T-MAC [Dam and Langendoen, 2003] are specially designed for WSN. B-MAC [Polastre et al., 2004] provides interfaces for application to control the backoff time when it initializes sending packet or when it detects collisions. S-MAC [Ye et al., 2002] and T-MAC [Dam and Langendoen, 2003] also use adaptive listening to reduce latency, and App-MAC [Du et al., 2005] uses the contention-based protocol to report event information. In App-MAC, they design a mechanisms to reduce collisions, e.g., filtering out some events with specified priority and some motes of specified types. They also design the CS assignment algorithms and the distributed CS competing algorithm to reduce collisions and save energy. Finally, variable length event data is considered in the design.

TDMA [Arisha et al., 2002] is a reservation-based protocol. In TDMA, a time period is divided into frames that provide each node with a transmission slot over which it can transmit data without collisions. NAMA [Bao and J.J.Garcia-Luna-Aceves, 2001] and TRAMA [Rajendran et al., 2003] are based on a distributed contention resolution algorithm that operates at each node based on the list of direct contenders and indirect interferences. They introduce energy-efficient collision-free channel access in WSN. GAMA-PS [Muir and Garcia-Luna-Aceves, 1998] and CRAMA-NTG [Garces and Garcia-Luna-Aceves, 1997a] use dynamic reservation to build a transmission schedule. IEEE 802.15.4 [IEEE802.15.4/D18, 2003] (Zigbee) designs a superframe structure, which combines an active part and an optional inactive part. The active part is further divided into a contention access period (CAP) and an optional contention free period (CFP).

3.8 Related Deceptive Data Detection Protocols

A previous effort that is close to our idea is presented to detect and diagnose data inconsistency failure in wireless sensor networks [Ssu et al., 2006]. The basic idea of their approach is to build node-disjoint path and use majority voting to detect inconsistency among collected data; however, in their work, they neither define data inconsistency models, nor propose a general framework for sensing systems.

In traditional systems, outlier detection problems have been explored. Three fundamental approaches have been proposed to detect outliers [Hodge and Austin, 2004]. First, unsupervised clustering techniques are used to determine the outliers with no prior knowledge of the data. They assume that errors or faults are separated from normal data and will thus appear as outliers. The basic idea is to classify data into different clusters and detect the data outside of the cluster as outliers. We can take advantage of those algorithms in our protocol. The other approach is called supervised classification, in which they model both normality and abnormality. If new data is classified to abnormal area, it will be an outlier. Those kinds of approaches require pre-knowledge so that they are not applicable in some highly dynamic and unpredictable environment. Another approach is a mixture of the previous two. It models only normality and a few cases abnormality.

A spatial outlier detection approach is proposed in [Shekhar et al., 2003]. They first formally model the problem of spatial outlier detection problem, and then they use a set of neighborhood aggregation function and distributive aggregation function to detect the outliers. Those aggregation functions can be used in our system as well. Adam, Janeja and Atluti design an algorithm to detect anomalies based on neighborhood information [Adam et al., 2004]. They explore both spatial and semantic relationship among the objects taking into consideration the underlying spatial process and the features of these spatial differentiated objects. Thus, they generate a set of both micro and macro neighborhoods, based on which outliers are detected. Bayesian network is a common used approach to classify sensor nodes according to the spatio-temporal correlations between those sensor nodes. For example, Bayesian belief networks are used in outlier detection [Janakiram et al., 2006].

A set of sensing data cleaning approaches have been proposed. Elnahrawy and Nath propose an approach for modeling and online learning of spatio-temporal correlations in sensing systems to detect and filter noise data [Elnahrawy and Nath, 2003]. This work shares the similar goal with ours, however, our approach is different from theirs by leveraging the consistency-driven concept and data dynamics for deceptive data detection and filtering. Their approach of outliers discovery complements our approach very well, and could be integrated into our system. A weighted moving average-based approach is proposed to clean sensor data [Zhuang et al., 2007]. The basic idea is to average the temporal-spatial data in an efficiency way, and detect noisy data based on the calculated average. This idea is very similar to the role-based deceptive data detection in our framework, but their approach is good to work in a relatively low dynamic environment while our approach can work in a highly dynamic environment in that we define importance of roles in different ways than them. Furthermore, our approach can take advantage of physical features in the physical system so that our approach can reach a confirmation in a fast and confident way.

Ye and *et al* propose to mechanism to statistically En-route filter injected false data [Ye et al., 2004, Ye et al., 2005]. They assume an event will be detected by multiple sensors and rely on the collective decisions of multiple sensors for false report detection. This approach is a security based approach and it does not really check the value of the collected data, in other words, this approach works at the event level but not at single data level. Declarative support for sensor data cleaning [Jeffery et al., 2006] is proposed by Jeffery *et al.* In their work, the authors utilize the temporal and spatial nature of sensor data to drive many of its cleaning process. As a result, they propose a framework named Extensible Sensor stream Processing (ESP) to segment the cleaning process into five programmable stages. Those stages are further integrated into a SQL-like language to clean the data. Compared with ESP, our framework for deceptive data detection and filters are more general by considering different types of deceptive data.

Several error correction algorithms have been presented in literature. A local, distributed algorithm to detect measurement error and infer missing readings in environmental applications is presented in [Mukhopadhyay et al., 2004b, Mukhopadhyay et al., 2004a]. In their work, a data distribution model is built based on the history data. Error is detected when there is a mismatch between the model prediction and the sensor reading. Sensor self-diagnostics or sensing reading calibration have been studied in [Li et al., 2007], in which preliminary data checking, analysis and sensor diagnosis are performed on-board. They try to build a set of rule-set based on data collected by different types of sensors and use those rule-set to analyze sensing data. Their goal is to detect fault readings at individual sensors. In this case, no malicious sensors are considered.

There are also deceptive data detection algorithms in vehicular networks. Probabilistic validation of aggregated data [Picconi et al., 2006] shows a way to probabilistically detect malicious cars that generate false aggregated information. In particular, they focus on validating speed and location information. To achieve this goal, they use a temper-proof service in the car as a proxy to buffer and transmit data. When data is received, a probabilistic verification is executed to validate received data. In general they adopt a security based approach to detect malicious aggregated data. Golle *et al* propose a mechanism to detect and correct malicious data in vehicular networks. They define a model of VANET, which specifies what events or sets of events are possible. A function maps a set of events to two values, valid and invalid. Only when the set of events are consistent, they are valid. Otherwise, malicious data is detected. How to check the consistent among the events is defined by a set of rules. This work shares similar ideas as our role-based deceptive data detection protocol in VANET, but we have different ways to detect deceptive data.

From above, we find that most outlier detection and other related algorithms assume a spatio-temporal distribution of the analyzed data, while our role-based deceptive data detection protocol does not assume any distribution on the sensing data. On the other hand, we propose a general framework to attack deceptive data detection and filtering problem. Furthermore, our protocol takes advantage of physical features in the cyber-physical systems such as vehicular networks.

CHAPTER 4

ORCHIS: A CONSISTENCY-DRIVEN DATA QUALITY MANAGEMENT FRAMEWORK

In this project, we intend to manage the quality of the collected data in an efficient way. Basically, we focus on two aspects in wireless sensing system design. One is consistency-driven data quality management. The other is energy efficient system design of wireless sensing systems, which is always one of the major considerations in wireless sensor system design. Thus, to integrate data consistency requirements and system support for energy-efficiency in wireless sensing system design, we propose Orchis [Sha and Shi, 2006a, Sha et al., 2008b, Sha and Shi, 2008], a consistency-driven energy efficient data quality management framework, which can be an architecture component that can be fitted in the four major architectures proposed in wireless sensor research. Our orchis framework consists of six major components, including a set of data consistency models customized to wireless sensing systems, a set of APIs to management the quality of collected data, a consistency-oriented data characteristics analysis to understand the features of sensing data, an adaptive protocol for data sampling, a set of consistency-driven cross layer protocols to support achieving the goals of data consistency and energy efficiency, as shown in Figure 9.1. Furthermore, all these protocols proposed here should be energy efficient. Thus, a metric named the lifetime of wireless sensor network is formally define, against which the energy efficiency property can be evaluated systematically from a system level view. Next we give an overview of each component one by one.

First of all, similar to that in traditional distributed systems, it is necessary to define a set of consistency models to evaluate the quality of collected data. With these consistency models, data consistency can be checked both at a single node and at a cluster head or base station after a series of data are collected. Moreover, data consistency models should be application-specific and take into consideration the special characteristics of wireless sensors as abstracted in Section 7.1, such as external inference, constrained resources, and unreliable wireless communication, and environment noises. These consistency models are the core of our framework. The application data quality requirement are represented by those consistency models and the quality of the collected data are evaluated based on the consistency models. Moreover, the parameters of the system protocols will be adjusted based on the parameters in consistency models.

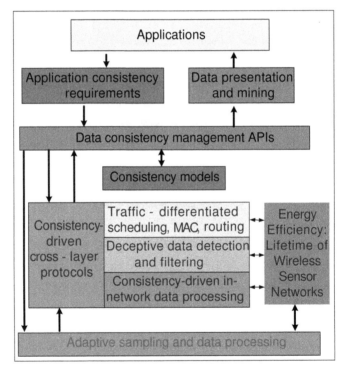

Figure 4.1: An overview of the Orchis framework.

Second, having the set of consistency models, we need to develop a set of APIs to manage data consistency. These APIs will be a bridge to connect data consistency models and application requirements from application scientists, as well as lower layer system protocols that collect sensing data from sensors. In other words, these APIs are essential and enable application scientists to disseminate and adjust consistent requirements, check consistency status, manage consistency requirements reactively, and detect and filter deceptive data. These APIs are connected with defined consistency models so that data quality can be evaluated. Also, these APIs provide interfaces for lower layer data collection protocols to efficiently transfer data to the sink, and present the collected data the application scientists.

Third, it is crucial to carefully analyze the collected sensing data, which not only helps us understand the features of monitored field, but also unveil any limitations and opportunities that should be considered in sensing system protocol design. Because in the extended sensing systems, after the traditional physical system are tightly entangled with the sensing systems, the system protocol design should integrate the special physical features of the systems. In this dissertation, we take an initial step and analyze one-month sensing data collected from a real-world water system surveillance application, focusing on data quality related data characteristics, such as the data similarity, data abnormality and failure patterns. The results from consistency-oriented sensing data analysis will be considered in the system protocol design.

Forth, due to the various data dynamics in different wireless sensor applications, we have to either collect a large amount of data, which is energy inefficient, or devise an adaptive protocol to improve the quality of collected data and take advantage of data consistency by considering data dynamics. Moreover, learned from the sensing data analysis, we find that there are high repentance in the sensing data, thus the quality of the collected data will not be affected too much if we do not deliver the repeated data. In this dissertation, we propose a data sampling and collection protocol that automatically adapts the data sampling rate according to data dynamics in the data field. This adaptive protocol can largely improve data quality and save energy. Furthermore, the zoom-in feature of the adaptive protocol helps us examine the collected data in detail thus to detect deceptive data and improve the quality of sensed data significantly.

Fifth, a set of consistency-driven cross-layer protocols are needed to support the goal of both collecting high data quality and achieving energy efficiency. Diverse data consistency requirements, various data dynamics, and changing data traffic resulting from adaptive protocols make it difficult to deliver all the messages timely and at the same time save energy. We design a suite of cross-layer protocols that allow the system to filter unnecessary data, sleep as much as possible, control the amount of traffic, and route packet while keep consistency, including adaptive resource allocation, duty cycle scheduling, and traffic-differentiated MAC and routing protocols, to support the cross-layer design. Among those protocols, in this dissertation, we take a deep step to explore the deceptive data detection protocol, because it plays a very important role in the data quality management. To detect and filter the deceptive data, we propose a framework, which includes several protocols to detect and filter two types of deceptive data, redundant data and false data. Especially, we give the detail of a role-differentiated deceptive data detection protocol.

Finally, although in the extended sensing systems, the constraints on the power supply are not as severe as those in traditional sensing systems. Energy efficiency is still one of the major design goals of any sensing systems as denoted in this dissertation, however, it lacks a good performance metric to evaluate the performance of energy efficiency property of the proposed protocols. Thus, except the five components introduced above, we also formally define a metric to evaluate the energy efficiency property of the proposed protocols. We model the lifetime of a wireless sensor network as an application specific concept, based on energy by considering the relationship between individual sensors and the whole sensor network, the importance of different sensors based on their positions, the link quality in transmission, and the connectivity and coverage of the sensor network. We use our lifetime models in this dissertation to evaluate the energy efficient performance of some proposed protocols.

In our Orchis framework [Sha and Shi, 2006a,Sha et al., 2008b,Sha and Shi, 2008], data consistency is controlled in the following ways. At first the consistency requirements are set by application scientists, which are translated into different consistency models. Then, data are collected by those proposed lower layer system protocols, where data are examined locally at each sensor node and on the path to the sink according to the specified consistency models, thus, some deceptive data are detected and filtered before they arrive at the sink. Due to the limited computation capability at individual sensor node, simple data consistency checking algorithms will be applied at node level consistency checking. When the sensing data are received at the sink, the consistency is checked again based on selected consistency models. Usually, the sink nodes are very powerful in terms of computation capability and they normally have more information than individual sensor nodes, thus computation intensive consistency checking algorithms can be applied to check the consistency of the collected data in a systemic way. Each time after consistency is checked, the system will make a decision. If the application finds that the consistency is satisfactory, it will continue to use the current parameter in all protocols. Otherwise, a modification to protocol parameters will be enforced to all the sensors through the consistency management APIs. In this way, the management of data quality forms a close loop feedback control and data quality can be bounded by consistency models or application requirements.

Last but not least, we discuss the function deployment of Orchis. In the Tenet architecture [Gnawali et al., 2006], researchers from University of Southern California and University of California at Los Angeles observed that the tiered principle is the best way to organize the sensor network. We believe that this hierarchical architecture has been very popular in the wireless sensing system applications. Orchis fits this new architecture very well, because consistency checking may require intensive computation and large storage, thus should be executed at the masters level. In addition, the Orchis framework can be easily applied in

Vehicular Networks and healthcare personal area networks as well. In Vehicular Networks, the consistency checking can be executed at each vehicle as well as at RSU based on different application requirements. For example, in the application of using warning message to improve the driving safety, data consistency checking has to be executed both at at individual vehicles and RSUs. Especially, the consistency checking at each vehicle is critical. While in some other applications, such as collect road traffic information, the consistency has to be checked at RSUs or even a remote server, where global information are necessary to generate a useful traffic information for some area. In healthcare personal-area sensing systems, the consistency can be checked at several layers as well. At the sensor layer, a local storage will be created to temporally store the recent collected data. Simple consistency checking algorithms can be applied to check that set of data. When the data are periodically transferred to a local sink, such as a smartphone or a PDA, consistency can be checked based on some more sophisticated algorithm.

In summery, our framework for data quality management is a general framework that can be applied in different kinds of wireless sensing system applications. In other words, different applications can choose suitable consistency models according to the application consistency requirements, from a set of very useful but distinct consistency models proposed in this dissertation. In this chapter, we just give a overview of our framework. The details of each component in our framework are introduced one by one in the following chapters.

CHAPTER 5

MODELING THE LIFETIME OF SENSING SYSTEMS

As specified in the last chapter, our goal in this dissertation is to extend the lifetime of the whole sensor network by considering data consistency. First, we need to define a metric to evaluate the performance of the energy efficiency. Here, we model the lifetime of wireless sensor networks. *The lifetime of wireless sensor network* is an application-specific, flexible concept. However, we can abstract and define *a remaining lifetime of wireless sensor network* first, which is defined as the weighted sum of *the lifetime of individual sensor* of all the sensors in the sensor network. Given that, we can define the the lifetime of the whole sensor network for three major application categories: active query, event-driven, and passive monitoring.

In an active query like applications, the lifetime of the whole sensor network can be defined as the maximum number of queries the sensor network can handle before the sensor network terminates. For an event-driven application, the the lifetime of the whole sensor network can be defined as total number of events the sensor network can process before the termination of it. For passive monitoring, the the lifetime of the whole sensor network can be defined as the total amount of time before the sensor network terminates. The termination of the sensor network is defined as the time when the remaining lifetime of wireless sensor network starts to keep unchanged that implies that the sensor network loses connectivity or the number of sensors with zero remaining lifetime exceeds a threshold which means that the sensor network becomes useless.*

Because the remaining lifetime of the whole sensor network is defined based on the remaining lifetime of sensors in the sensor network. We define the remaining lifetime of

---

*Here, we assume the energy consumption of regular maintenance overhead is negligible, and will be considered later.

the single sensor first, then model the remaining lifetime of the whole sensor network, and finally define the lifetime of the whole sensor network.

5.1 Assumptions and Definitions of Parameters

Several assumptions made by our model are listed here. All the symbols used in this analysis are listed in Table 5.1 as well.

- All the sensors are homogeneous, i.e., the physical capacity of each sensor is same.

- The location information is available, either by physical devices such as GPS or by topology discovery algorithms [Deb et al., 2002, Haeberlen et al., 2004, Moore et al., 2004].

- The sensors in the sensor network are almost evenly distributed and dense enough to cover the whole area.

- The location of each sensor is stationary.

- The sensor's power is limited that it can only communicate with its neighbors within its communication range. Multi-hop is required to communicate with others that is outside communication range.

- The data sink is fixed, which is usually true in the real deployment.

5.2 Definition of Remaining Lifetime of Sensors

Remaining Lifetime of Individual Sensor is defined as the remaining normalized energy of the sensor at some moment, N_m. Here we normalize the remaining initial energy of all the sensors in the sensor network as one. In the time during which a query is executed, the energy is consumed when the sensor receives or sends a query message and a reply message.

Variables	Description
ϵ_{jiq}	In the ith query the amount of energy consumed when one query message go through the jth sensor, specifically $\epsilon_{jiq,rcv}$ for receiving and $\epsilon_{jiq,snd}$ for sending. When we assume all the query messages are same, ϵ_{jiq} can be reduced to ϵ_q
ϵ_{jir}	In the ith query the amount of energy consumed when one reply message go through the jth sensor, specifically $\epsilon_{jir,rcv}$ for receiving and $\epsilon_{jir,snd}$ for sending. When we assume all the reply messages are same, ϵ_{jir} can be reduced to ϵ_r
E_j	The total initial energy of jth sensor. When we assume all the sensors to be homogeneous, E_j is equal to E.
$\frac{1}{f}$	An application-specific parameter to determine the possibility of a sensor generating a reply to a query
N_{jiq}	The number of the query messages that go through the jth sensor during the ith query
N_{jir}	The number of the reply messages that go through the jth sensor during the ith query
r	The communication range of each sensor
S_q,S_r	The size of the query message and reply message separately
P_{jiq}	The probability that the ith query message go through the jth sensor node
P_{jir}	The probability that the reply messages for ith query goes through the jth sensor node
P_{ji}	The probability that the query message or reply message will go through the jth sensor node during ith query
$P(B)_{ij}$	The probability that the sensor farther than the jth sensor to the sink is selected as the destination in ith query
$P(A)_{ij}$	The probability that the message going through the jth sensor in ith query when that message goes through the circle area where the jth sensor node is located
N_{far}	The number of the sensor nodes that are farther than the jth sensor node to the sink
N_{total}, N_n	The number of the total sensor nodes in the sensor network
N_{nbrs}	The number of the sensors in the communication range
N_{fail}	The number of the sensors that are out of power
θ	The maximum number of depleted sensors when the sensor network's functionality is correct
ρ	The density of the sensor nodes in the sensor network
d_{jis}	The distance from the jth sensor node to the sink (or delegate) during the ith query
N_m	The sequence number of queries processed in active query, the sequence number of events handled in event-driven and the amount of time used in passive monitoring
w_j	The weight(importance) of the jth sensor node in the sensor network
d_{max}	The ratio of the maximum distance between every two sensor nodes in the sensor network to the communication range r
$\mathcal{L}$	The remaining lifetime of the whole sensor network
$L(j)$	The remaining lifetime of the jth sensor
LFT	The lifetime of the whole sensor network

Table 5.1: A list of variables used in Chapter 5.

So the remaining lifetime of the sensor is the total energy of each sensor minus the energy consumed when the messages go through the sensor. It can be defined as

$$L(j) = 1 - \sum_{i=1}^{N_m} \frac{\epsilon_{jiq} * N_{jiq} + \epsilon_{jir} * N_{jir}}{E_j}$$

We borrow the same energy model and symbols used in [Heinzelman et al., 2000] to calculate energy consumption of each message transmission. The energy consumed when the sensor receives a message of size k is

$$\epsilon_{rcv} = \epsilon_{elec} * k$$

and the energy consumed on sending a message of size k is

$$\epsilon_{snd} = \epsilon_{elec} * k + \epsilon_{amp} * r^2 * k$$

So, we have

$$\epsilon_{jiq} = \epsilon_{jiq,rcv} + \epsilon_{jiq,snd}$$

and

$$\epsilon_{jir} = \epsilon_{jir,rcv} + \epsilon_{jir,snd}$$

5.3 Probability of ith Message Through the jth Sensor

To calculate the $L(j)$, we should calculate N_{jiq} and N_{jir} first, which is related to the probability of the message go through the jth sensor in the ith query. Thus we need to define the probability of the ith message going through the jth sensor first. As we can observe, the query messages directed to the sensors far away from the sink and reply messages from far away sensors to the sink will both go through some sensors nearer to the sink than these sensors. While the messages from or to the sensors near to the sink will not go through the farther ones. Thus if all sensors have the same probability to be the query destination, the probability that the message go through the nearby sensors of the sink is larger than that of going through the far away sensors to the sink. Figure 5.1 shows the

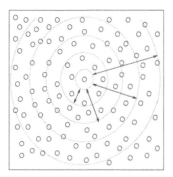

Figure 5.1: Message propagation in sensor networks.

message propagation in the sensor network from a macro view. A message is routed from the sink to the jth sensor step by step, at each step the message must be in one circle area as shown in the figure. If there are n sensors in one circle area, the probability that the message is hold by one specified sensor in that circle area is $\frac{1}{n}$. In point to point routing, we can assume that the probability for ith query message go through the jth sensor is the same as the probability for the reply messages for that query go through the jth sensor, i.e., $P_{jiq} = P_{jir}$. Here we use P_{ji} as the probability that the message will go through the jth sensor in the ith query. While in the broadcast, $P_{jiq} = 1$, for every sensor will get the query.

Let A denote the event that the message goes through the jth sensor when it goes through the circle area the node locates, and B be the event that the destination of the message i is farther to the sink than the jth node. A and B are two independent events. Let $P(B)_{ij}$ be the probability of B, and $P(A)_{ij}$ be the probability of A. Then the probability that the ith message will go through the jth node is

$$P_{ij} = P(A|B)_{ij} = P(A)_{ij}P(B)_{ij}$$

Here,

$$P(B)_{ij} = \frac{N_{far}}{N_{total}} = \frac{N_n - \pi d_{jis}{}^2 \rho}{N_n}$$

and

$$
\begin{aligned}
P(A)_{ij} &= \frac{1}{\pi(((\lfloor \frac{d_{jis}}{r} \rfloor + 1) * r)^2 - (\lfloor \frac{d_{jis}}{r} \rfloor * r)^2)\rho} \\
&= \frac{1}{\pi(2\lfloor \frac{d_{jis}}{r} \rfloor + 1)r^2 \rho}
\end{aligned}
$$

Thus

$$P_{ij} = \frac{1}{\pi(2\lfloor \frac{d_{jis}}{r} \rfloor + 1)r^2 \rho} \frac{N_n - \pi d_{jis}{}^2 \rho}{N_n}$$

Now we can calculate the remaining lifetime of sensor in the case of unicast in `Traditional` and broadcast in both `Traditional` and `IQ`. In above formula, the difference between `Traditional` and `IQ` lies in d_{jis}, which is constant in `Traditional` for each sensor because the sink is fixed all the time and is changing in `IQ` because of the shifting of the delegate.

5.4 Remaining Lifetime of Sensors in Unicast using Traditional

In the case of unicast, the message will go through only the sensors on the path between the sink and the destination once for each query, so does the reply message for that query. Thus the remaining lifetime of each sensor in unicast using `Traditional` can be defined as

$$
\begin{aligned}
L(j) &= 1 - \sum_{i=1}^{N_m} \frac{\epsilon_{jiq} N_{jiq} + \epsilon_{jir} N_{jir}}{E_j} \\
&= 1 - \sum_{i=1}^{N_m} \frac{\epsilon_{jiq} * P_{ji} * 1 + \epsilon_{jir} * P_{ji} * 1}{E_j}
\end{aligned}
$$

Because we assume the sensors are homogeneous and the same type messages are having the same size, E_j is equal to E, $\epsilon_{jiq} = \epsilon_q$ and $\epsilon_{jir} = \epsilon_r$, the remaining lifetime of each sensor in the point to point routing is

$$L(j) = 1 - \frac{1}{E} \frac{(\epsilon_q + \epsilon_r) N_m}{\pi(2\lfloor \frac{d_{jis}}{r} \rfloor + 1)r^2 \rho} \frac{N_n - \pi d_{jis}^2 \rho}{N_n}$$

5.5 Remaining Lifetime of Sensors in Broadcast using Traditional

Next, we analyze the remaining lifetime of each sensor when the query message is broadcasted to all the sensors from the sink. In this case, the query message floods to all the sensors while the reply messages will go through the sensors on its path to the sink. For each query message if we assume the probability that one sensor will generate a reply message is $\frac{1}{f}$, the remaining lifetime of each sensor in broadcast using Traditional is

$$
\begin{aligned}
L(j) &= 1 - \sum_{i=1}^{N_m} \frac{\epsilon_{jiq} * N_{jiq} + \epsilon_{jir} * N_{jir}}{E_j} \\
&= 1 - \sum_{i=1}^{N_m} \frac{\epsilon_{jiq} * N_{nbrs} + \epsilon_{jir} P_{jir} N_n \frac{1}{f}}{E_j}
\end{aligned}
$$

Because the query is routed by flood, i.e., each sensor will get a query from its neighbors and send a query to its neighbors. So the remaining lifetime of sensor in the broadcast traditional query is

$$
L(j) = 1 - \frac{N_m \epsilon_q N_{nbrs}}{E} - \frac{\epsilon_r N_m}{E f \pi r^2 \rho} \frac{N_n - \pi d_{jis}^2 \rho}{2 \lfloor \frac{d_{jis}}{r} \rfloor + 1}
$$

5.6 Remaining Lifetime of Sensors in IQ

In the indirect query, a query is directed to a randomly selected delegate, then the delegate acts as the sink to take care of query forwarding, data collection, and data transmitting back to the sink. If the probability of each sensor to be a delegate is the same, when there are totally N_m queries been processed, and there are totally N_n sensors in the sensor network, for each sensor the possible times it is selected as a delegate is $\frac{N_m}{N_n}$. The number of times that the sensor is located in the area with $d_{jis} = kr$ to the delegate is the number of the sensors located in the circle area between kr and $kr + 1$ to that sensor. The number of the sensors located in the circle area between kr and $kr+1$ is $\pi(2k+1)r^2\rho$. Thus the remaining lifetime of each sensor in IQ can be defined as

$$
\begin{aligned}
L(j) &= 1 - \sum_{i=1}^{N_m} \frac{\epsilon_{jiq} * N_{jiq} + \epsilon_{jir} * N_{jir}}{E_j} \\
&= 1 - \frac{\epsilon_q N_m N_{nbrs}}{E} - \frac{1}{E}\frac{N_n}{f}\sum_{i=1}^{N_m}\epsilon_r N_{jir} - \frac{N_m}{N_n}\frac{N_n}{fE}\epsilon_{r,rcv} \\
&= 1 - \frac{\epsilon_q N_m N_{nbrs}}{E} - \frac{1}{E}\frac{N_n}{f}\sum_{k=1}^{d_{max}}(\epsilon_r \frac{N_m}{N_n} * P_{jis}(\pi\rho(((k+1)r)^2 - (kr)^2))) \\
&\quad - \frac{N_m}{fE}\epsilon_{r,rcv} \\
&= 1 - \frac{\epsilon_q N_m N_{nbrs}}{E} - \frac{\epsilon_r N_m d_{max}}{Ef} + \frac{\epsilon_r N_m}{Ef}\frac{\pi\rho r^2}{2N_n}d_{max}(d_{max}+1) - \frac{N_m}{fE}\epsilon_{r,rcv}
\end{aligned}
$$

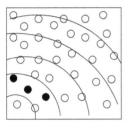

(a) The sensors near the sink are dead

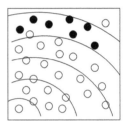

(b) the sonsors far from the sink are dead

Figure 5.2: An example of the importance of different sensors, assuming the data sink is located at the low-left corner.

Based on the remaining lifetime of sensors, we define the remaining lifetime of the whole sensor network in the following section. First we define the importance of sensors.

5.7 Importance of Different Sensor Nodes

The failure[†] of sensors will cause the sensor network to act improperly, but the level of the damage it causes is different, which is the reason why we think the previous definition

---

[†]In this dissertation, we consider only the failure resulted from the depletion of energy.

of the lifetime as the time before the first sensor failure or first message failure without considering the location of the sensor is unsatisfactory. For the same number of failure sensors, the damage may be very slight at sometime and the sensor network still performs almost normally, while sometimes it may be very serious which makes the sensor network lose its most functionality. Two cases are described in the Figure 5.2(a) and (b) as an example. In the figure, the black nodes represent the sensors that have run out of energy and the white ones denote the ones that are still alive. In both Figure 5.2(a) and (b) the sensor networks cannot act as it suppose to do since in both cases the sensor network cannot gain data from some sensors, but in Figure 5.2(a), although there are only three failed sensors, the sink cannot get data from most of the sensors. In Figure 5.2(b), there are seven dead nodes, but the sink can still get data from most of the sensors in the sensor network. So the damage to the sensor network by the failure sensors is not only related to the number of failed sensors but also related to the location of the failed sensors. To this end, sensors in the sensor network have different importance. We define a factor named *weight* for each sensor to count the importance of that sensor. Based on above analysis, the nearer the sensor to the sink, the more important it is. So we define the weight of each sensor as following:

$$w_j = c \frac{1}{d_{jis}^2}$$

Here c is a constant.

5.8 Remaining Lifetime of the Whole Sensor Network

In Section 5.2, 5.4, 5.5, 5.6, we have defined the remaining lifetime for sensors. Now we are in a position to examine the remaining lifetime of the whole sensor network. We consider *the remaining lifetime of the whole sensor network* as the sum of the weighted remaining lifetime of all sensors in the sensor network. Thus the remaining lifetime of the whole sensor network is

$$\mathcal{L} = \sum_{j=1}^{N_n} w_j L(j)$$

From this definition, we can easily get the remaining lifetime of the whole sensor network in unicast, broadcast using **Traditional** and in **IQ**. Specifically, in unicast, the remaining lifetime of the whole sensor networks is defined as:

$$\mathcal{L} = \sum_{j=1}^{N_n} c\frac{1}{d_{jis}^2}(1 - \frac{1}{E}\frac{(\epsilon_q+\epsilon_r)N_m}{\pi(2\lfloor\frac{d_{jis}}{r}\rfloor+1)r^2\rho}\frac{N_n-\pi d_{jis}^2\rho}{N_n})$$
$$\approx c\pi\rho(2\ln(d_{max}) + \frac{\pi^2}{6}) - \frac{c(\epsilon_q+\epsilon_r)N_m\pi^2}{6Er^2} + \frac{c(\epsilon_q+\epsilon_r)N_m}{Er^2 d_{max}}$$

In the case of broadcast using **Traditional**, the remaining lifetime of the whole sensor network is defined as:

$$\mathcal{L} = \sum_{j=1}^{N_n} c\frac{1}{d_{jis}^2}(1 - \frac{N_m\epsilon_q N_{nbrs}}{E} - \frac{\epsilon_r N_m}{Ef\pi r^2\rho}\frac{N_n-\pi d_{jis}^2\rho}{2\lfloor\frac{d_{jis}}{r}\rfloor+1})$$
$$\approx c\pi\rho(2\ln d_{max} + \frac{\pi^2}{6})(1 - \frac{\epsilon_q N_m N_{nbrs}}{E}) - \frac{c\pi^2\epsilon_r N_m N_n}{6Efr^2} + \frac{c\epsilon_r N_m \pi\rho d_{max}}{Ef}$$

Similarly in **IQ**, the remaining lifetime of the whole sensor network is defined as:

$$\mathcal{L} = \sum_{j=1}^{N_n} c\frac{1}{d_{jis}^2}(1 - \frac{\epsilon_q N_m N_{nbrs}}{E} - \frac{\epsilon_r N_m d_{max}}{Ef}$$
$$+ \frac{\epsilon_r N_n}{Ef}\frac{\pi\rho r^2}{2N_n}d_{max}(d_{max}+1) - \frac{N_m}{Ef}\epsilon_{r,rcv})$$
$$\approx c\pi\rho(1 - \frac{\epsilon_q N_m N_{nbrs}}{E} - \frac{\epsilon_r N_m d_{max}}{Ef}$$
$$+ \frac{\epsilon_r N_m}{Ef}\frac{\pi\rho r^2}{2N_n}*d_{max}(d_{max}+1) - \frac{N_m}{fE}\epsilon_{r,rcv})(2\ln d_{max} + \frac{\pi^2}{6})$$

Based on the remaining lifetime of the whole sensor network, the lifetime of the sensor network can be formally defined as:

$$\text{LFT} = \{ N_m \mid \mathcal{L}(N_m - 1) < \mathcal{L}(N_m) \ \& \ \mathcal{L}(N_m + 1) = \mathcal{L}(N_m) \text{ or } N_{fail} \geq \theta \}$$

where θ is a pre-defined threshold of maximum number of the failure sensors in the sensor network and N_{fail} is the number of the failure sensors.

Based on these models, we depict the detailed analysis of different query protocols in the following section.

Query Types	Unicast (Traditional)	Broadcast (Traditional)	IQ
$0 < d < r$	$1 - \frac{158 N_m}{10^6}$	$1 - \frac{8280 N_m}{10^6}$	$1 - \frac{2149 N_m}{10^6}$
$d = 7r$	$1 - \frac{8 N_m}{10^6}$	$1 - \frac{2000 N_m}{10^6}$	$1 - \frac{2149 N_m}{10^6}$
$d = 14r$	1	$1 - \frac{1670 N_m}{10^6}$	$1 - \frac{2149 N_m}{10^6}$

Table 5.2: Comparison of the remaining lifetime of different nodes in different locations.

5.9 Analytical Comparison: Traditional vs. IQ

One of the goals of modeling is to evaluate the performance of different protocols. Now we are in the position to compare Traditional with IQ in terms of the remaining lifetime of sensors and the remaining lifetime of the whole sensor network.

To quantitatively compare these two query protocols, we adopt the practical values of sensor parameters obtained from Berkeley motes [Hill et al., 2000], including the initial energy and the energy consumption rate. In [Hill et al., 2000] two $1.5V$ batteries rated at $575mAh$ are used for each sensor, so the initial total energy of each sensor is $1.725J$. The energy to transmit and receive a single bit is $1\mu J$ and $0.5\mu J$ respectively. We assume the size of query message and reply message to be $240bits$ and $1200bits$ separately. Thus it takes $240\mu J$ to transmit a query message and $120\mu J$ to receive a query message; and it takes $1200\mu J$ and $600\mu J$ to transmit and receive a reply message. If we assume the total number of sensors in the sensor network is 1500 and the density of the sensor network is 1 per $1000m^2$, the maximum distance between every two sensors is 14r, where r is the communication range equal to $50m$. We also assume the probability of one sensor generating a corresponding reply message is $1/30$, thus $f = 30$.

First we compare the sensors located at different areas in the sensor network based on the remaining lifetime of the sensors. Table 5.2 depicts the remaining lifetime of the sensors located at different regions in the sensor field in the context of different query protocols. Here we select the sensors with its distance to the sink as within one communication range, with seven times communication range, and with 14 times communication range respectively.

Query Types	Unicast (Traditional)	Broadcast (Traditional)	IQ
Remaining lifetime	$\frac{21733-0.8N_m}{10^6}$	$\frac{21733-69N_m}{10^6}$	$\frac{21733-46N_m}{10^6}$

Table 5.3: Comparison of the remaining lifetime of the whole sensor network.

The last range is the largest distance to the sink. From the figures in the table, we can find that the remaining lifetime of the sensor increases along with the increase of the distance between the sensor to the sink in Traditional, but it increases quicker in the broadcast communication than in unicast communication, because the load imbalance is accumulated faster in broadcast case. Thus the sensors near to the sink will consume a lot of energy and fail very quickly, which results in the earlier termination of the whole sensor network. From this observation, we find that the unbalanced load results in the short lifetime of the sensor network. On the contrary, as we expected, the remaining lifetime of sensors located in different regions using the IQ protocol are almost same, which denotes that IQ indeed does a good job on balancing the load among all sensors.

We also compare the remaining lifetime of the sensor network here. The results of comparison of three types of query protocol are listed in Table 5.3. From these deduced results, we can find that the sensor network using the IQ protocol has longer remaining lifetime than that of using Traditional by providing a global optimization to balance the load to the whole sensor network. Furthermore, considering these two tables together, in broadcast using Traditional, when $N_m = 121$, although there are still a large amount of energy ($\frac{13384}{10^6}$ from Table 5.3) remaining in the sensor network, it will never be used because the sensor network is down when all sensors within the sink's communication range are down (see in Table 5.2 when $N_m = 121$). While in IQ, because the load is balanced, no sensors will run out of energy much earlier than others. So most energy of each sensor

Variables	Values
Communication Range	30m
Number of Nodes	400
Total Energy of Each Sensor	3 Joules
Packet Size	240, 1200 bits
Energy Dissipated for Receiving	50 nJ/bit
Energy Dissipated for Transmission	50 nJ/bit
Energy Dissipated for Transmit Amplifier	100 pJ/bit/m^2
Bandwidth	40kbps

Table 5.4: Simulation parameters.

will be effectively used in IQ. To this end, we think that IQ is more energy efficient than Traditional, i.e, the energy utilization is much lower in IQ than that in Traditional.

From the above analysis, we conclude that IQ extends the lifetime of the whole sensor network because it balances the load to all the sensors in the sensor network, which again validates our argument that the load balance plays a very important role in the lifetime of the whole sensor network.

5.10 Simulation Verification

To verify the analytical results, we conduct a detailed simulation using the Capricorn [Sha et al., 2004], a large-scale discrete-event driven simulator. In our simulation, 400 nodes are scattered to a $600m \times 600m$ square field. We use the GPSR routing protocol implemented in routing layer of the simulator to deliver message. All simulation parameters are listed in Table 5.4. In this section, we evaluate Traditional and IQ in terms of energy consumption, RLSN and LSN.

5.10.1 Energy Consumption

First we compare energy consumption of each sensor in Traditional and IQ, which can reflect RLIS. Figure 5.3 reports the energy consumption of sensors in two protocols after

400 queries have been processed, where x-axis and y-axis together decide the location of each sensor node and z-axis depicts the value of energy consumption. Figure 5.3(a) shows

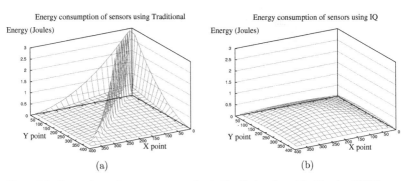

(a)   (b)

Figure 5.3: Comparison of energy consumption in `Traditional` and in `IQ` using GPSR.

that some sensors in `Traditional` consume a lot of energy, especially those located along the two edges and the diagonal line of the sensor field to which the data sink belongs. So these sensors are energy hungry which consume all 3 Joules, while sensors located outside this region just consume as little as 0.1 Joules after 400 queries. Obviously, the energy consumption in `Traditional` is very unbalanced. On the contrary, the load in `IQ` balances very well, as shown in Figure 5.3(b), where there are no energy intensive nodes. In `IQ` the maximum energy consumption is 0.6 J and the minimum energy consumption is 0.04 J. In other words, in the `IQ` protocol, by running out of the total 3 J, the sensor network can process at least 2000 queries.

5.10.2 Simulation results of RLSN

In this section, we compare the two protocols in terms of RLSN. According to the default values of simulation parameters, the initial RLSN is 21, which is calculated from the formula in Section 7.2. Figure 5.4 shows the simulation results, where x-axis is the number of queries, and y-axis represents RLSN.

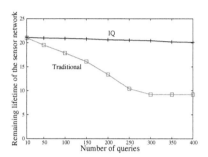

Figure 5.4: Comparison of RLSN by using Traditional and IQ.

From the figure, we find that RLSN decreases along with the increase of the number of processed queries. In Traditional, RLSN drops very quickly from 21 to less than 10 during 300 queries have been processed. After 300 queries have been processed, RLSN using Traditional keeps stable. This is because the sensor network is already dead after 300 queries. In other words, no more query message can be sent from the sink to other alive sensors, thus no more energy is consumed. This does not mean that the remaining energy in the alive sensors is saved. On the contrary, these energy is just wasted, which can never be used in the future. In case of IQ, RLSN drops very slowly and smoothly. After 400 queries have been processed, RLSN is still good using IQ. A lot of energy is saved to be used for the future queries. From this point of view, we argue that IQ is more energy efficient than Traditional.

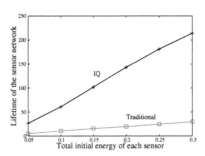

Figure 5.5: Comparison of LSN by using `Traditional` and `IQ`.

5.10.3 Simulation results of LSN

Finally, we compare LSN, which is more interesting to application scientists and system designers. We set the value of θ (the threshold to determine the aliveness of the sensor network) to 90%. The comparison between `Traditional` and `IQ` protocols is reported in Figure 5.5, where x-axis is the initial energy of each sensor and y-axis is the value of LSN. From the figure, it can be easily seen that LSN increases almost linearly with the increase of initial energy in both `Traditional` and `IQ`. However, LSN increases much faster in `IQ` than that in `Traditional`, where the lifetime in the `Traditional` model is about $\frac{1}{6}$ of that in `IQ`. Additionally, if we increase the value of θ, the gap between the `Traditional` and `IQ` will become much larger. Thus we conclude that `IQ` indeed extends LSN several times as that of `Traditional`.

In this chapter, we formally define the remaining lifetime of individual sensor, the remaining lifetime of sensor networks, and the lifetime of wireless sensor network, which is an application specific performance metric to evaluate the performance of the proposed protocol. Based on these models, we compare two query protocols; both theoretical and simulation results show that `IQ` balances the load so that extends the lifetime of the sensor network. Having this system level performance metric to evaluate energy efficiency, we are ready to provide an energy efficient system design, which is denoted in the next several chapters.

CHAPTER 6

QUALITY-ORIENTED SENSING DADA ANALYSIS

With the increasing number of deployments of sensor systems, in which the main function is to collect interesting data at the sink, it is becoming crucial to carefully analyze the large amount of collected data. However, this problem is neglected in previous research, which mainly focuses on energy efficient, reliable sensor systems design and optimization. Although data quality management attracts more and more attention in the last two years [Li et al., 2006, Sha and Shi, 2008], proposing novel data quality management mechanisms is still an important and interesting research topic. We argue that sensor system optimization and data quality management are closely related to the characteristics of collected data, in other words, sensor system optimization and data quality management should take data characteristics into consideration. Thus, in this dissertation, we take an initial step to characterize the data quality using a set of one-month data collected by a real-world water system surveillance application. The data set consists of water level, precipitation, and gauge voltage measurements from 13 gauges located around Lake Winnebago, St. Clair River and Detroit River in January 2008.

Our data analysis focuses on *quality oriented data analysis*. In quality oriented data analysis, we intend to discover two types of data, namely similarity data and abnormal data. The significance of our discovery is two-fold. On one hand, it helps us understand the laws and changes in the monitored field. On the other hand, it unveils the limitations in the current sensor system design, and provides us with a strong ground upon which we can base our future WSN systems design.

Our study reveals several interesting facts. First, information similarity, including pattern similarity and numerical similarity, is very common, which provides a good opportunity to trade off energy efficiency and data quality. Second, different parameters exhibit different data characteristics, which suggests that adaptive protocols using variable sampling rates can bring in significant improvements. Third, spatial correlation analysis and multi-modality correlation analysis provide a way to evaluate data integrity and to detect conflicting data that usually indicates appearances of sensor malfunction or interesting events. Fourth, abnormal data may appear all the time, and continuous appearance of abnormal data usually suggests a failure or an interesting event. Finally, external harsh environmental conditions may be the most important factor on inflicting failures in environmental applications. Despite these failures and possible sensing failures, data quality is still satisfactory.

6.1 Background

The United States Army Corps of Engineers (USACE) in Detroit District, has 22 data collection platforms commonly referred to as sensor nodes or gauges, deployed around the St. Clair and Detroit rivers in southeast Michigan as well as the Lake Winnebago watershed southwest of Green Bay, Wisconsin. One month data in January 2008 from 13 of the 22 gauges were made available for this study. Each sensor node collects battery voltage, water level and precipitation except the Dunn Paper gauge (G1) which collects battery voltage, air temperature and water temperature. However, precipitation data for the St. Clair/Detroit river system is not used in this work, because that "data" in the raw files is simply an artifact of the gauge programming. For convenience, we name each sensor node as G1, where 'G' stands for "gauge." The gauge locations of G1, G2 and G3 are on the St. Clair River, G4 is located on the Detroit River, and G5 through G13 are spread around the Lake Winnebago watershed. Gauges G5 through G13 are shown in Figure 6.1, and the other remaining gauges are shown in Figure 6.2. It is worth mentioning that G8 and G10 suffered many failures throughout the period of study. Therefore, any data analysis on G8 or G10 will mostly look like "weird" or at least different from the other gauges.

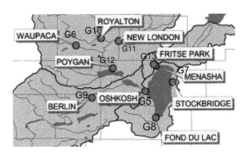

Figure 6.1: Lake Winnebago Watershed.

Data samples are sent from each gauge to the GOES satellite, once every hour or every four hours, depending on whether the station has a high baud rate transmitter or not. High baud rate transmitters send data every hour. Data is then sent from the satellite to a central location in Wallops Island in Virginia, where the data samples are collected and arranged in files for later download through a regular ftp service. We conducted our analysis directly using the un-decoded files. This raw data set has not been subject to any quality control procedure, and thus provide a good opportunity to study false readings happening in sensing systems. Water level and precipitation are sampled once every hour, whereas voltage is sampled once every hour or every two hours. Water level is measured against the IGLD Datum 1985, which works as the base to measure current water level. So negative water level means it is below the local IGLD Datum 1985, though that does not happen often. Precipitation data is supposed to be constantly increasing (except when the gauge resets as part of its normal operation). To figure out how much precipitation fell over a one-hour period, the difference between consecutive samples are calculated and reported. The measurements for voltage and precipitation are in volts and inches respectively. Water

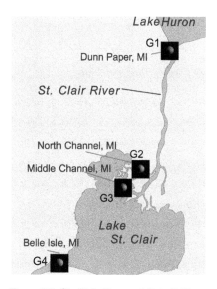

Figure 6.2: St. Clair River and Detroit River.

level is reported in meters, centimeters and feet. For convenience, we converted the readings
for water level in meters.

Based on the raw data, we focus on individual time series analysis for each sensor
and one of its measure attribute, correlation between different data series. For each data
series, we investigated information redundancy pattern and proposed methods to detect
abnormal data. The redundancy pattern is analyzed in terms of pattern similarity and
numerical similarity. Correlation between data series is examined against different attributes
for the same sensor node and the same attribute for different sensor nodes. The relation
between water level and precipitation is carefully studied with different endurance interval
settings. As to spatial correlation, we looked into the similarity of water level collected
by different sensor nodes, and the similarity of precipitation collected by different sensor
nodes. In addition, to detect abnormal water level, we analyzed empirical distribution for

the difference of scaled water level. The water level is scaled according to our statistical analysis.

6.2 Quality-Oriented Sensing Data Analysis

The goal of data analysis is to understand and discover the rules and patterns in the data set, as well as to detect interesting events in the monitored field. Also, we argue that the implication of the data analysis can unveil the limitations in the current wireless sensor network designs. In this section, we focus on quality-oriented sensing data analysis. How to define data quality is still an open problem. Here, We define high quality data as the data that contains the most information from the monitored field.

We analyze the data collected from an environmental monitoring application, which uses 13 gauges to monitor water level, accumulative participation and voltage of in Lake Winnebago, St. Clair River, Detroit River during the entire month of January 2008. To understand the quality of the collected data, we try to discover the spatial and temporal relationship among those data; specifically, we are mostly interested in detecting two types of data, redundant data and abnormal data. Usually, redundant data, which we name as similarity, will not affect the overall quality of the collected data when it is removed. Contrary to redundant data, abnormal data, which largely affects the data quality, should be examined more carefully, because it usually denotes sensor failures, malicious attacks or interesting events. In the rest of this section, we first examine each sensor parameter individually from the temporal perspective to discover any pattern and numerical similarity, as well as abnormal data. Then, we check the relationship between two parameters, water level and accumulated precipitation. Finally, we figure out the relationship between same type sensors located in different locations.

6.2.1   Time Series Analysis for Individual Parameter

For each individual parameter, we define two types of similarity, *the pattern similarity* and *the numerical similarity*. Here, we define a pattern as the continuous reappearance of the same value sensed by one sensor, and the number of continuous reappearance is called pattern length. Note that a pattern must have a minimum length of 2. Thus, each pattern is a two tuple $< key, length >$. For example, if the sensor reads a series of $4, 4, 4, 5, 5, 4, 5$, we detect the patterns as $< 4, 3 >$ and $< 5, 2 >$, and the number of appearance of each pattern is 1. We use the pattern reappearance ratio, which is the ratio of the pattern data in the whole data, to measure the pattern similarity among the data. The numerical similarity records the number of reappearance of the same numerical value. For example, if a sensor reads a series of $4, 4, 4, 5, 5, 4, 5$, we get numerical similarity as 4 times of appearance of value 4 and 3 times of appearance of value 5. Similarly, the numerical similarity ratio is used to evaluate the value similarity, which is defined as the ratio of the reappeared sensor readings in all sensor readings.

Pattern Similarity

We detect all patterns for all monitoring parameters in 13 gauges. Here, we pick up gauge G5 as a typical example to show the patterns we detected as well as reappearance times of the pattern, which is shown in Figure 6.3.

From the figure, we do find specific patterns in the collected data, and the number of total patterns is small for all three parameters. Water level of gauge G5 has the largest number of patterns, which is 33, whereas, precipitation has the smallest number of patterns, which is 19. Some patterns have very large pattern length. For example, the largest pattern length for water level and for precipitation are 77 and 139 respectively. This indicates that water level and precipitation values stay constant for a long period of time at the area where G5 located. The number of appearances of each pattern is mostly small, especially for patterns with large length. This is because we set up endurance interval as $[0.00, 0.00]$, thus, very small difference between two keys, such as 14.66 and 14.67, are distinguish. Here endurance interval is an interval within which the difference between two readings can be ignored, for example, if the endurance interval is $[-0.02, 0.02]$, 14.67 and 14.66 can be regarded as the same. Because of unavoidable system error in measurement and applications lowered requirements on accuracy, it is reasonable to set up an endurance interval for each monitoring parameter. Another reason is that we define different patterns even when they have the same key value but different lengths.

Figure 6.4 shows the pattern reappearance ratio, where "-" means there is no available data. In the figure, we find that voltage has the smallest pattern reappearance ratio, which suggests that the changes of the voltage are very frequent. This is also because we distinguished the pattern keys in extremely fine granularity; however, even in such a fine granularity, both patterns in water level and precipitation show a large ratio of pattern similarity. For example, the smallest pattern reappearance ratio is 0.43 in G3, and the largest pattern reappearance ratio is 0.94 in G12. While precipitation shows the largest pattern similarity, which can be seen not only from the least number of patterns in Figure 6.3, but also from the fact that it has all pattern reappearance ratio larger than 0.77; actually, most pattern reappearance ratio of precipitation is about 0.99 for all gauges. We can expect precipitation to stay stable at most time. It may change suddenly, however, after this sudden change, it goes back to normal and stabilizes for a long period of time. Voltage has the most varying pattern reappearance ratios, which ranges from 0.04 to 0.96, showing that the performance of the power supply is really independent and highly dynamic. The goal of the sensor network applications is to collect meaningful data, thus, most of those applications can endure a certain level of data inaccuracy, which will not affect our discovery of the rules and events in the monitoring field. We reexamine the pattern reappearance ratio after we lower the accuracy requirements on the collected data and set up different endurance intervals for three parameters. The resulted pattern reappearance ratio is depicted in Figure 6.5, where the three numbers under the title are the endurance intervals, which are mostly 10% of possible largest changes, i.e., we allow voltage to endure 0.2 volts changes, water level to endure 0.04 meter changes, and precipitation to endure 0.04 inch changes. Note that different units are used for water level and precipitation, i.e., meter for water level and inch for precipitation, which we keep the original units as in the raw data.

Comparing Figure 6.5 to Figure 6.4, we can find that almost all pattern reappearance ratios increased by increasing the endurance interval, especially for those with small reappearance ratio in Figure 6.4. After we increase the endurance interval, we can see that 50% of the voltage data pattern reappearance ratio is larger than 0.95, while water level and precipitation pattern reappearance ratio do not change too much compared to that in voltage; however, most of them are still larger than those in Figure 6.4. From both figures, we can see that there is a big pattern reappearance ratio.

In our definition, pattern length means the number of continuous appearance of the same sensor reading. Thus, we try to figure out the distribution of the pattern length in terms of variable endurance interval, as shown in Figure 6.6, Figure 6.7, and Figure 6.8, where the x-axis is the length of the pattern and the y-axis is the CDF of the pattern length. From the figure, we find that most patterns have short patten length. For example, when the endurance interval is set to be [0.00, 0.00], 90% of voltage patterns have length less than 10, and about 70% of water level patterns and about 60% of precipitation patterns have length less than 10. However, different parameters have different pattern lengths. In the figure, we can see that voltage, which has almost all pattern length less than 20, has more short length patterns than water level and precipitation, while precipitation has the longest length among the three parameters, where about 30% of the precipitation pattern has length longer than 20. This observation shows that precipitation is stable at most of the time, but the reading of the voltage has high dynamics. By increasing the endurance interval, more patterns have longer length appear. For example, when water level endurance interval is increased to [−0.04, 0.04], more than 30% of the patterns have length between 140 to 180.

Numerical Similarity

Having studied pattern similarity, we move on to check the numerical similarity. Numerical similarity focuses on the numerical value reappearance of the sensing data, which differs from pattern similarity in that numerical similarity does not intend to detect any pattern. For the numerical similarity, we identify the number of appearance for each individual value. Figure 6.9, Figure 6.10 and Figure 6.11 shows the numerical distribution of the collected data, where the x-axis is the numerical value of the sensing reading and the y-axis denotes the number of appearance of the corresponding numerical value. Note that we pick up the data collected by gauge G5 as an example.

V Pattern	Appearances	W Pattern	Appearances	P Pattern	Appearances
<14.09, 2>	1	<0.67, 2>	1	<1.67, 2>	1
<14.58, 2>	1	<0.68, 2>	1	<1.77, 2>	1
<14.59, 2>	1	<0.69, 2>	2	<2.28, 2>	1
<14.63, 2>	2	<0.70, 2>	1	<2.50, 2>	1
<14.64, 2>	1	<0.71, 2>	2	<2.52, 2>	1
<14.66, 2>	2	<0.72, 2>	1	<2.30, 5>	1
<14.67, 2>	2	<0.68, 3>	1	<2.47, 5>	1
<14.68, 2>	3	<0.69, 3>	2	<2.24, 7>	1
<14.69, 2>	1	<0.70, 3>	2	<2.34, 7>	1
<14.71, 2>	1	<0.71, 3>	4	<2.40, 8>	1
<14.73, 2>	1	<0.72, 3>	2	<1.69, 16>	1
<14.57, 3>	1	<0.70, 4>	2	<2.36, 22>	1
<14.60, 3>	1	<0.71, 4>	1	<2.20, 38>	1
<14.67, 3>	3	<0.72, 4>	3	<2.36, 44>	1
<14.72, 3>	1	<0.69, 5>	2	<2.11, 49>	1
<14.65, 4>	1	<0.70, 5>	4	<2.53, 67>	1
<14.67, 4>	2	<0.70, 9>	1	<2.41, 97>	1
<14.68, 4>	3	<0.72, 9>	1	<2.51, 102>	1
<14.69, 4>	2	<0.71, 11>	1	<1.29, 139>	1
<14.61, 5>	1	<0.71, 14>	1		
<14.64, 5>	1	<0.68, 16>	1		
<14.67, 5>	1	<0.71, 16>	1		
<14.68, 5>	2	<0.71, 18>	1		
<14.68, 6>	1	<0.71, 22>	1		
<14.70, 6>	1	<0.67, 28>	1		
<14.67, 7>	2	<0.68, 29>	1		
<14.69, 7>	1	<0.71, 30>	1		
<14.68, 8>	2	<0.70, 34>	1		
<14.68, 9>	2	<0.70, 38>	1		
<14.68, 10>	1	<0.69, 45>	1		
<14.68, 11>	1	<0.69, 47>	1		
<14.67, 12>	1	<0.68, 69>	1		
		<0.72, 77>	1		

Figure 6.3: Detected patterns and the number of appearance in gauge G5.

Gauge ID	Voltage Pattern Reappearance Ratio	Water Level Pattern Reappearance Ratio	Precipitation Pattern Reappearance Ratio
G1	0.07	-	-
G2	0.85	0.64	-
G3	0.78	0.43	-
G4	0.38	0.57	-
G5	0.50	0.88	0.92
G6	0.17	0.84	0.89
G7	0.96	0.69	0.93
G8	-	-	-
G9	0.23	0.87	0.89
G10	0.04	0.44	0.77
G11	0.10	0.83	0.90
G12	0.87	0.94	0.91
G13	0.05	0.88	0.89

Figure 6.4: Pattern reappearance ratio with zero endurance interval.

Gauge ID	Voltage Pattern Reappearance Ratio [-0.1, 0.1]	Water Level Pattern Reappearance Ratio [-0.02, 0.02]	Precipitation Pattern Reappearance Ratio [-0.02, 0.02]
G1	0.59	-	-
G2	1.00	0.85	-
G3	1.00	0.74	-
G4	0.96	0.78	-
G5	0.88	0.99	0.95
G6	0.98	0.94	0.92
G7	1.00	0.99	0.95
G8	-	-	-
G9	1.00	0.94	0.92
G10	0.64	0.72	0.79
G11	0.78	0.93	0.94
G12	1.00	0.99	0.94
G13	0.80	0.95	0.94

Figure 6.5: Pattern reappearance ratio with increased endure interval.

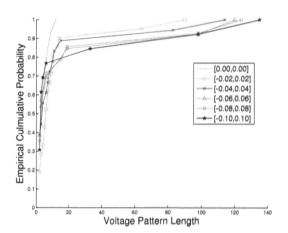

Figure 6.6: CDF of pattern length of Voltage.

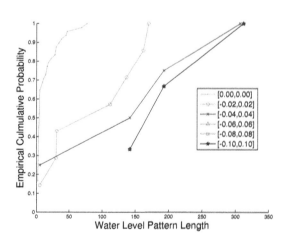

Figure 6.7: CDF of pattern length: of water level.

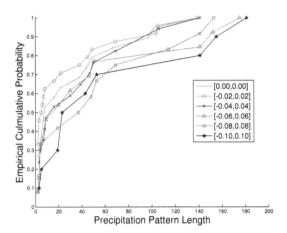

Figure 6.8: CDF of pattern length of precipitation.

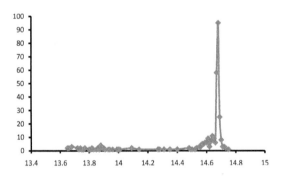

Figure 6.9: The number of appearances for each numerical values of voltage.

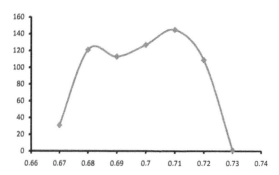

Figure 6.10: The number of appearances for each numerical values of water level.

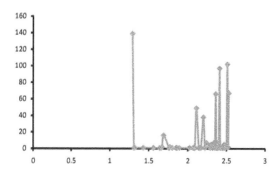

Figure 6.11: The number of appearances for each numerical values of precipitation.

In the figure, we find that those three parameters exhibit totally different distributions. The reading of the voltage and water level are very close to normal distribution with $\mu = 14.22, \sigma = 0.38$ and $\mu = 0.7, sigma = 0.02$ respectively. The voltage readings are more centralized to value 14.7, while water level readings are more broadly distributed from 0.68 to 0.72 and centralized at 0.71. The reading of precipitation shows no obvious distribution. It spreads from about 1.25 to 2.55. Some precipitation values such as 1.29 and 2.51, appear much more times than others, which means no rain or snow falls for a long time after the precipitation value is read, while other precipitation readings only appear several times, which mainly depicts some transitional states during a continuous rain or snow falling. Like the pattern similarity ratio, numerical reappearance ratio is used to evaluate the numerical similarity.

Gauge ID	Voltage Numerical Reappearance Ratio	Water Level Numerical Reappearance Ratio	Precipitation Numerical Reappearance Ratio
G1	0.84	-	-
G2	0.99	0.64	-
G3	0.97	0.43	-
G4	0.91	0.57	-
G5	0.82	0.88	0.92
G6	0.94	0.84	0.89
G7	0.99	0.69	0.93
G8	-	-	-
G9	0.97	0.87	0.89
G10	0.77	0.44	0.77
G11	0.82	0.83	0.90
G12	0.99	0.94	0.91
G13	0.89	0.88	0.89

Figure 6.12: Numerical reappearance ratio with endurance interval [0.00, 0.00].

Figure 6.12 presents the numerical reappearance ratio of the sensing data at all gauges. We can see that all parameters exhibit very high reappearance ratio. Compared to pattern reappearance ratio, numerical reappearance ratio is much larger for voltage, fairly larger for water level, and comparable for precipitation. For example, the voltage pattern reappearance ratio in G1, G10 and G11 is less than 10%, while the voltage numerical pattern reappearance is close to 80%. The large difference implies that although the numerical readings of the voltage have a large similarity, they fluctuate very frequently and there are no obvious patterns in voltage readings. The two reappearance ratios of precipitation do

Gauge ID	Voltage Numerical Reappearance Ratio [-0.1, 0.1]	Water Level Numerical Reappearance Ratio [-0.02, 0.02]	Precipitation Numerical Reappearance Ratio [-0.02, 0.02]
G1	0.95	-	-
G2	0.99	0.97	-
G3	0.99	0.94	-
G4	0.97	0.95	-
G5	0.94	0.99	0.95
G6	0.98	0.98	0.93
G7	1.00	1.00	0.95
G8	-	-	-
G9	0.99	0.96	0.93
G10	0.89	0.93	0.90
G11	0.92	0.96	0.94
G12	1.00	0.99	0.95
G13	0.95	0.97	0.94

Figure 6.13: Numerical reappearance ratio with increased endurance interval.

not differ too much, which suggests that reappearance patterns play an important role in precipitation.

Similar to what we have done in the pattern similarity analysis, we increase the endurance interval to a certain level. Here, we set the endurance interval to the same value as we did in the pattern similarity analysis. As a result, most numerical appearance ratios are increased by increasing the endurance interval; however, the increasing rate is not as big as the one in pattern similarity analysis. From Figure 6.13, we really find that the numerical redundancy is very high in all three types of sensing data. For instance, after we increase the endurance interval, the numerical reappearance ratio is mostly over 90%. We also try to mine the pattern of the data change in terms of the time series. We calculate the coefficiency in the time series with different time periods such as 24 hours, 48 hours and so on, however, we find that all the coefficiencies are very low, thus, we believe that there is no strong clues showing the periodically reappearance pattern in data changes.

Abnormal Data Detection

Abnormal data may result from sensor malfunction, data loss during the communication, faked data inserted by malicious nodes, or the appearance of an interesting event. We try to detect abnormal data based on the presented numerical value of the data. Basically, two types of abnormal data can be detected. One is the out-of-range data, and the other is dramatic changing data.

Gauge ID	Parameter	Position	Value
G6	Water Level	Reading # 45	62.79
G10	Voltage	Reading #47	1.00
G10	Voltage	Reading #119	1.00
G10	Voltage	Reading #126	1.00
G10	Voltage	Reading #127	1.00
G10	Voltage	Reading #323	1.00

Figure 6.14: Detected out-of-range readings.

Figure 6.14 shows the appearance of the out-of-range data, which is the data out of the possible valid range defined by the domain scientists. Based on the figure, we figure out that most sensing data are within the normal range. We find out-of-range data only at two gauges, G6 and G10, and G6 only has one invalid reading. Considering the failure patterns, we find that G10 has a maximum number of failures as well. So, we believe there are some relations between the probability of abnormal readings and the probability of failures.

Figure 6.15 explains hourly water level changes in gauge G3, where we find that their distributions are close to normal distribution based on normal probability plot, which is a graphical technique for assessing whether or not a data set is approximately normally distributed. For such data, 3-sigma limits is a common practice to base the control limit, i.e., whenever a data point falls out of 3 times the standard deviation from its average value, it is assumed that the process is probably out of control. In the figure, two horizontal lines depict the upper and lower 3-sigma limits. We find that only several points are out of the two limits, which means domain scientists do not need to check the cause of water level changes at most time. The similar patterns are detected in all gauges as shown in Figure 6.16, where most gauges have water level changes within 3-sigma limits. Investigation is deserved when out-of-limit changes are detected to find the cause of the abnormality.

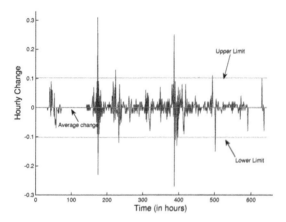

Figure 6.15: Limit control of G3's hourly water level changes.

G2	G3	G4	G5	G6	G7	G9	G10	G11	G12	G13
2.17%	1.55%	2.95%	0.00%	0.31%	0.16%	0.00%	1.40%	0.16%	6.06%	1.86%

Figure 6.16: Out-of-Limit ratio for hourly water level changes.

Implications

Learned from above similarity and abnormal analysis, we argue that we need to revisit system protocol design by integrating the intrinsic features of the monitoring parameters.

First, we can take advantage of the large amount of data similarity. Because data similarity is common, it is not necessary to transfer all the collected data to the gateway. Quality-assured local data processing, aggregation and compression algorithms are necessary to remove redundant data and reduce overall data volume but keep the quality of the collected data at a satisfactory level. By enduring a certain level of data inaccuracy, we can reduce the total amount of collected data up to 90% according to the pattern and numerical reappearance ratios. In addition, strong patterns are helpful to estimate the future data and detect abnormal data.

Second, we can use different data sampling rates for different monitoring parameters. For example, we discover that the changes in voltage is much more frequent than those in precipitation. Thus, we need to increase the sampling rate to sense voltage data more frequently, whereas, decrease the sampling rate for precipitation. Furthermore, in the sensor readings for precipitation, some of them reappear a large amount of times, while others only appear once. Usually, the readings that only appear once or twice imply a highly dynamic environment. Therefore, it is better to increase sampling rate so that we can detect the details in changes.

Third, there may be a lot of abnormal data existing in the sensor reading. Basically, they can be classified to two categories. One type is transitional, which disappears very quickly. We can mostly ignore this type of data without affecting overall data quality by replacing it with a reasonable value. The other type is continuous, which typically lasts a longer period of time. This type of abnormal data usually implies malicious data or interesting events. When continuous abnormal data is detected, more attention should be paid to them at the early stage. For example, more data should be sampled and reported to the gateway as fast as possible.

Finally, various data sampling rates may result in different amount of data traffic. Samplings for different parameters and detected abnormal data may have different priorities in their delivery to the gateway. A well designed data collection protocol is necessary to achieve this goal.

6.2.2 Multi-Modality and Spatial Sensing Data Analysis

In the last subsection, we analyzed the similarity and abnormal data for each parameter individually. In this subsection, we analyze the relationship between two types of sensing data, water level and precipitation. Moreover, we try to explore the spatial relationship at different locations.

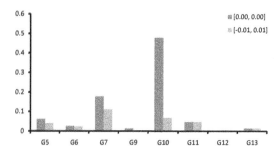

Figure 6.17: Conflict ratio of water level and precipitation.

Although water level can be affected by many factors, including moisture when it starts raining, rainfall intensity, and even temperature and slope of the land, we believe that there is a relationship between water level and precipitation. Mostly when precipitation increases, water level should also increase. We count the ratio of the conflict, which is defined as the appearance when precipitation increases but water level decreases, to verify this relationship. Figure 6.17 records the conflict ratio between water level and precipitation. In the figure, the x-axis depicts the gauge ID, and the y-axis shows the conflict ratio. The dark blue bar and the gray bar denote the conflict ratio with endurance interval $[0.00, 0.00]$ and $[-0.01, 0.01]$ independently. From the figure we observe that in most cases the conflict ratio is less than

6%, which verifies that water level is closely related to precipitation; however, there are two gauges with conflicts larger than 10%, i.e., G7 has conflict ratio of 18% and G10 has conflict ratio of 48%. After a carefully examination, we figure out that G10's high conflict ratio is related with lots of failures it has. While G7's high conflict ratio may be caused by other reasons, because when precipitation increases only a little, other factors, such as moisture and temperature, may play major roles to determine water level. This is verified by the fact that when we increase the endurance interval a little, the conflict ratio decreases very fast, and it eventually disappears when we set endurance interval to $[-0.01, 0.01]$ for water level and $[-0.02, 0.02]$ for precipitation.

Water Level	G2	G3	G4	G5	G6	G7	G9	G11	G12	G13
G2	1	0.924	0.174	0.344	0.083	0.314	0.64	0.471	0.513	0.569
G3	0.924	1	0.306	0.428	0.254	0.377	0.601	0.505	0.525	0.591
G4	0.174	0.306	1	0.486	0.413	0.35	0.35	0.54	0.436	0.528
G5	0.344	0.428	0.486	1	0.159	0.916	0.699	0.888	0.875	0.555
G6	0.083	0.254	0.413	0.159	1	0.113	-0.1	0.002	-0.01	0.32
G7	0.314	0.377	0.35	0.916	0.113	1	0.659	0.832	0.848	0.474
G9	0.64	0.601	0.35	0.699	-0.1	0.659	1	0.825	0.835	0.7
G11	0.471	0.505	0.54	0.888	0.002	0.832	0.825	1	0.909	0.627
G12	0.513	0.525	0.436	0.875	-0.01	0.848	0.835	0.909	1	0.652
G13	0.569	0.591	0.528	0.555	0.32	0.474	0.7	0.627	0.652	1

Figure 6.18: Spatial correlation of water level.

Precipitation	G5	G6	G7	G9	G11	G12	G13
G5	1	0.977	0.996	0.995	0.985	0.996	0.996
G6	0.977	1	0.959	0.99	0.996	0.967	0.968
G7	0.996	0.959	1	0.983	0.969	0.999	0.998
G9	0.995	0.99	0.983	1	0.992	0.988	0.988
G11	0.985	0.996	0.969	0.992	1	0.975	0.975
G12	0.996	0.967	0.999	0.988	0.975	1	0.999
G13	0.996	0.968	0.998	0.988	0.975	0.999	1

Figure 6.19: Spatial correlation of precipitation.

We analyze spatial correlation for all of the three parameters. Because there are no direct communications among sensors at different locations in this application, we do not expect spatial correlation among voltage readings at the different gauges, which is validated by the collected data. The calculated co-efficiency value between any two gauges is less than 0.54 and 99% of them is less than 0.32. However, we do find some spatial correlation for both water level and precipitation based on the data sensed from various gauges. The results are depicted in Figure 6.18 and 6.19 retrospectively.

In the figure for precipitation, we only have data for listed gauges. We can see that all gauges with precipitation data have very large co-efficiency value because they are all located at LAKE WINNEBAGO, which means that the weather in that area is pretty uniform. When there is a rain fall at the location of one of the gauges, it is most probably raining at the locations of the other gauges as well. Water level also exhibits the similar pattern. In Figure 6.18, gauges located closely usually have high co-efficiency values, which results in similarity in water level changes, while gauges located far away usually have no obvious similarity in terms of water level changes. For instance, gauges can be grouped into several small groups with similar water level changes based on the calculated large co-efficiency values. Thus, gauge G2 and G3 are within one group with co-efficiency value larger than 92%. We can see that both of them are located in St. Clair River. G4 is the only gauge in Detroit River, so it has no high co-efficiency with any other gauges. Moreover, gauge G5, G7, G11, and G12 show high similarity because they are located closely. Thus, we believe that geographical similarity exists in the sensed data for water level and precipitation.

Implication

Multi-modality and spatial sensing data analysis helps us to find the correlation between different parameters and geological correlation of the same parameter. Therefore, data collected by the correlated sensors can be used as a reference to calibrate the sensing data. For example, an increase in precipitation mostly results in an increase in water level. When there are some conflicts between them, we need to take a close look and figure out the reason of the conflict. Furthermore, we can take advantage of similarity in different parameters or sensors located in different locations. Quality-assured aggregation can be applied in this scenario to reduce the volume of sensing data. Thus, multi-modality models and spatial models are very useful in quality-assured data collection protocol design.

CHAPTER 7

MODELING DATA CONSISTENCY IN SENSING SYSTEMS

Data consistency models, used as metrics to evaluate the quality of the collected data, are the core of the Orchis framework. Thus in this chapter, we formally define a set of data consistency models for different applications. We will first abstract the consistency related features of wireless sensor networks. Then, we give out the formal definition of our consistency models. Finally, we propose a set of APIs to manage the consistency.

7.1 Consistency Requirements Analysis

Data consistency is an important problem in computer architecture, distributed systems, database, and collaborative systems [Peterson and Davie, 2003,Ramakrishnan, 1998,Tanenbaum and van Steen, 2002]. A lot of consistency models have been proposed in these fields. However, these models are usually not applicable in WSN because of the specific characteristics of WSN. Thus, consistency models, the key to evaluate the quality of the collected data, should be remodelled for WSN applications. In this section, we first analyze the difference between WSNs and traditional distributed systems in terms of consistency; then, we abstract the data consistency requirements in WSN.

Although a WSN is an instance of a distributed system, there are several significant differences between WSNs and traditional distributed systems. First, WSNs are resource constrained systems. Due to the constraints of the memory and the large amount of the data, the data are usually not stored in sensors for a long period, but they will form data streams to be delivered to the sinks or base stations. As a result, data consistency in WSN will not focus on the read/write consistency among multiple data replicas as in traditional distributed systems; instead, data consistency in WSN is more interested in the spatial and temporal consistency of the same data, i.e., the consistency among several appearances of the data at different locations and in different time. Second, WSN applications may have more interests in a set of data which can depict the trends of the monitoring parameter or report an event by combining these data together. Thus, consistency models for data streams are more important than those for individual data. In this dissertation, consistency models for both types of data are modeled. Third, compared with traditional distributed systems, the unreliable wireless communication is common, rather than abnormal, in WSN. Although retransmission is a strategy to rectify the effect caused by the unreliable wireless communication, there is no simple technique that can guarantee the successful delivery of a message. Thus, in the consistency model, the data loss due to wireless communication should also be considered. Furthermore, in previous definition of the data consistency [Tanenbaum and van Steen, 2002], the effect of channel noises and intended attacks are neglected. We argue that attacks are normal nowadays, and the security technologies should be integrated in the system design to prevent attacks.

In summary, we conclude that consistency models in traditional distributed systems that basically discuss the read/write consistency among different replicas are not sufficient to be applied in WSNs. Given the specific features of resource constraints and unreliable communication, consistency models in WSN should be remodelled.

Considering both individual data and data streams, we argue that the quality of the data should be examined from three perspectives: *the numerical consistency, the temporal consistency, and the frequency consistency*, as shown in Figure 7.1. The *numerical consistency* requires that the collected data should be accurate. Here we have two kinds of concerns on numerical errors: *absolute* and *relative*. Absolute numerical error happens when the sensor reading is out of normal reading range, which can be pre-set by applications. In the case of absolute numerical error, we can remove it and estimate a reasonable value for it. Relative numerical error depicts the error between the real field reading and the corresponding data at the sink. To trade off the resource usage and data accuracy, we can leverage estimation

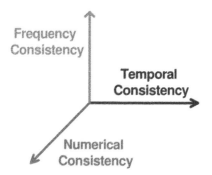

Figure 7.1: A three-dimension view of consistency requirements.

technologies to estimate readings at the sink while still providing the data with the same level of accuracy. As a result, some sensor readings can be dropped to save resource usage. Subsequently, there are relative numerical errors between the real ground truth and the collected data at the sink. The *temporal consistency* means that the data should be delivered to the sink before or by it is expected. The *frequency consistency* controls the frequency of dramatic data changes and abnormal readings of data streams, i.e., the number of dramatic data changes and the number of readings out of normal reading range in one time interval should be limited by the application specific consistency requirements. Given this definition, we can see that the quality of each individual data is determined by the numerical and temporal consistency, while the quality of data streams is depicted by the combination of three consistency perspectives.

All these three types of consistency are application-specific concepts, thus different application may have various consistency requirements for them respectively. For example, in a patient monitoring system, emergency conditions of a patient should be reported to the control panel or caregivers as fast as possible. Otherwise, the patient may be in a dangerous condition. Thus, most systems that need quick response or have high real-time requirements usually have high requirements on the temporal consistency. Other systems may have no strict time requirements on the collected data. For instance, a roadside monitoring system that counts the number of passed vehicles in one area may only need the data to be reported twice a day. In this case, data aggregation is more possible because some aggregation functions need to wait until sufficient data are available. However, these kinds of systems may have high accuracy requirements (i.e., numerical consistency requirement) on the collected data. Usually the applications that have high accuracy requirements may have strict requirements on high frequency requirements for the purpose of quick system reaction when some abnormal data is detected, e.g., some event-detection applications may care more on the dramatic data changes and abnormal readings, which usually represent the occurrence of some interesting events or attacks.

7.2   Consistency Models

We are in a position to propose data consistency models to evaluate the different data quality. In this section, we first define a general data format used in our models. Then, we model data consistency for individual data followed by data consistency models for data streams. Note that we assume that unnecessary data are detected and filtered by detecting protocols, and an estimation technique [Sha and Shi, 2006a] is used to fill these data at the sink. In this dissertation, we define consistency as follows. The destination and the intermediate sensors on the path from the data source to the destination always receive accurate or meaningful data in a timely fashion.

7.2.1   Data Format

Before we formally model data consistency, we define a general data format that will be used in consistency models. Considering both the temporal and numerical consistency requirements and noticing that the frequency consistency requirement is derived from the numerical values of the data, we define the data format used in consistency models as follows,

$$(p_i, T_{stamp}, Seq\_Ran, Val, ReT)$$

where p_i denotes that the data is from the ith sensor for parameter p; T_{stamp} specifies the time when the value is sampled and Seq_Ran is the range of the sequence number of the reading at the ith sensor for parameter p. Seq_Ran contains only one number where there is no two continuous readings that have the same value. Val is the value of the reading with sequential number in Seq_Ran, while ReT depicts the remaining time before it is expected by the sink. The initial value of ReT is set the same as temporal consistency requirements and the value of T_{stmap}, Seq_Ran, and Val are set locally by the ith node.

As we analyzed in the above subsection, different applications can have various consistency requirements. An example of a consistency requirement is given here, ($NE =$ $0.2, Max\_T = 3m, Max\_Diff = 1.2, Max\_CHG = 5, Range = [1,4], Max\_OutRange =$ 3), where $NE = 0.2$ means that the application can endure ± 0.2 numerical error; $Max\_T =$ $3m$ denotes the data should be received at sink in 3 minutes after it is sampled; $Max\_Diff =$ 1.2 and $Max\_CHG = 5$ together define that the number of dramatic changes, the difference between two continuous readings exceeds ± 1.2, should not exceed five; and $Range$ and $Max\_OutRange$ requires that the number of readings with value not in the $Range$ should be less than $Max\_OutRange$. These consistency requirements are pre-distributed by other protocols (Section 7.1). Based on these requirements, we can abstract two consistency semantics: (1) the difference between any two continuous meaningful readings (at the sink) should be larger than 0.2; and (2) if the number of dramatic changes in one minute exceeds five, the collected data is not good enough because the detail of the changing is ignored. Next, we formally model data consistency by using these abstracted semantics. General ReT should be larger than zero when the data arrives at the sink, all the collected data satisfy the consistency semantics of the application, specified by the combination of Val and Seq. The numerical consistency requirements are specified by the consistency semantics, e.g., the application can endure 0.2 error for temperature. In addition, in our consistency models, we assume the value of the reading is accurate and how to improve the quality of sensing data is beyond the scope of this dissertation.

7.2.2 Consistency Models for Individual Data

We consider consistency both for individual data and data stream. For each piece of data, we want to keep the corrected data fresh and accurate, so we will check both temporal consistency and numerical consistency. Moreover, the consistency can be checked at different locations and the data may be delivered by various protocols according to different application consistency requirements, so we model three types of data consistency, *the hop-consistency, the single-path consistency,* and *the multiple-path consistency.* The hop-consistency means that the data should keep consistency in each hop, while the single-path consistency and the multiple-path consistency imply that data consistency holds when the data is transmitted from the source to the sink using a single path and multiple paths respectively. *The hop-consistency* is checked at each hop when the data is transferred and it is maintained if the data packet still has sufficient time to be transferred from current node to the sink and the value of the new arrive reading is within the range of the consistency semantics. We define it as below,

$$HopConsist = (InSemantics(Val, Val_{last})$$
$$\&OnTime(ReT, EsT_t)) - - - (1)$$

where, $InSemantics$ judges the numerical consistency by checking whether the new arriving value and the cached last value follow the pre-defined consistency semantics, and $OnTime$ checks the temporal consistency, i.e., $ReT \geq EsT_t$ denotes that the node has enough time to deliver the data to the sink. Because the hop-consistency is checked at each hop along the path, so it is very useful to detect attacks on data and filter redundant aggregated data when data aggregation is applied by using estimation technologies. This will reduce the source usage while providing the same level of data accuracy.

The hop-consistency defines the consistency only at each hop, however, the end-to-end consistency between data source and data destination is more important from the viewpoint of applications. We define this type of consistency as the path-consistency, which is usually checked at the sink. According to the different routing strategies and application requirements, we define two types of path-consistency, the single-path consistency using single path routing and the multiple-path consistency using the multiple-path routing or flooding. The difference between them lies in that at most one copy of the same data is reported to the sink in the single-path consistency, while several typically copies of the same data will be received at the sink in the multiple-path consistency. Both types of path-consistency consider two concepts, temporal consistency and numerical consistency. We model the single-path consistency as below,

$$SPathConsist = (InSemantics(Val, Val_{last})$$
$$\&(ReT \geq 0)) --- (2)$$

Similar to functions in the hop-consistency, $InSemantics$ checks the numerical consistency in terms of the consistency semantics. The temporal consistency is reflected by the condition that ReT is larger than 0 when the data arrives at the sink. In the multiple-path consistency, several copies of same data will be delivered to the sink. Thus the sink will check the consistency as k-consistency, which means at least k copies of the same data should be reported to the sink in time. The multiple-path consistency modeled as below is very useful to detect the faked readings inserted by malicious nodes (i.e.,fault detection), which might be very important to some applcaitions.

$$MPathConsist = (InSemantics(Val, Val_{last})$$
$$\&ReT \geq 0\&Count(P_i, Seq\_Ran) \geq k) --- (3)$$

Compared with the single-path consistency, the multiple-path consistency has one more requirement about the number of copies for the same data, denoted by $Count(P_i, Seq\_Ran)$.

7.2.3 Consistency Models for Data Streams

In WSN, data are usually collected in the format of data streams. Individual data may not have significant meaning, while they are useful when the set of the data are considered together. Thus, we argue that consistency models for a set of data, data streams, denoted as $D = \{d_1, d_2, ..., d_n\}$, are the same important, if not more important, as the models for individual data. For data streams, we propose six types of consistency models to satisfy different consistency levels, including *the strict consistency, the α-loss consistency, the partial consistency, the trend consistency, the range frequency consistency* and *the change frequency consistency*. All these consistency models are taking into consideration of application requirements from three consistency perspectives as analyzed in Section 7.1. The first four consider the different levels of numerical and temporal consistency while the rest two focus on the frequency related consistency.

The strict consistency has the most strict requirements to the consistency of collected data, so it can be used in applications that have extremely high consistency requirements. To satisfy the strict consistency, three requirements must be satisfied. First, no data is missed during transmission, i.e., the packet with each sequence number i should be received at the sink. Second, the temporal consistency is satisfied, i.e., for all received data at the sink, $ReT_i \geq 0$. Third, the numerical consistency in terms of consistency semantics is maintained, e.g., any two continuous readings d_i and d_{i+1} in the data set D received at the sink, are out of each other's endurance range. So the restricted consistency is modeled as

$$StrictConsist = (InSemantics(D)\&\forall i ReT_i \geq 0$$
$$\&\forall i \in [1, n], d_i \in D) - - - (4)$$

The strict consistency differs from the hop-consistency because it is defined based on a set of data and requires no data lose, so it is stricter than the hop-consistency from this point of view. Not all applications require the strict consistency, which may be almost impossible to achieve in a wireless communication based system such as WSN. If we allow some data loss during transmission, we get the α-loss consistency, where all received data should keep temporal consistency and at least $1 - \alpha$ percent of totally sampled data should be received at the sink. So the α-loss consistency can be modeled as

$$\alpha - LossConsist = (InSemantics(D)\&\forall iReTi \geq 0$$
$$\&Counter(D) \geq (1 - \alpha) * \max(Seq\_Ran)) - - - (5)$$

where $InSemantics$ checks the numerical consistency as before. All the received data are temporal consistent and the number of total received data is large than $1 - \alpha$ percent of the number of total sampled data, which is checked based on the sequence number of the received data. For example,if totally n pieces of data should be received based on the value of Seq_Ran, and the real received number is $Counter(D)$, we can check if the condition in above formula is satisfied. The α-loss consistency is suitable for applications that have high real-time requirements. The value of α is adjustable to cater to the numerical consistency requirements of the applications.

In addition to releasing numerical consistency requirements, we can also release the temporal consistency requirements, which results in the partial consistency. In the partial consistency, not all the data are required and the temporal consistency are not so strict, thus it is modeled as

$$ParConsist = (InSemantics(D)\&\forall iReT_i \geq -a$$
$$\&Counter(D) \geq (1 - \alpha) * \max(Seq\_Ran)) - - - (6)$$

The partial consistency is similar to the α-loss consistency except that the temporal consistency requirement is released. This consistency model is useful in applications where aggregation applies, which have numerical consistency requirements but low temporal consistency requirements.

If we further release the numerical consistency requirement, we get another consistency model named the trend consistency, which is defined as follows,

$$TrendConsist = (TrendSatisfy(D)) --- (7)$$

where $TrandSatisfy$ detects if the trend of data streams is maintained. Mechanisms are needed to evaluate the valid trends. For instance, we might utilize some algorithms from the signal processing field to evaluate the quality of data streams, e.g., frequency domain features. This consistency model matches the trend requirement (Section 7.1) of some WSN applications very well, which could be used in attack-resilient data collection protocols.

Now we consider the abnormal data readings in data collection. In certain applications, the application scientists may have pre-knowledge of the normal data range of their application. This is very helpful to filter erroneous readings, which are resulted from a variety of reasons, including intended attacks. Also, if the number of abnormal readings exceeds a certain number pre-set by the application, the application scientists may need to check the abnormal phenomenon. The notification of the abnormal phenomenon will be triggered by a violation to the range frequency consistency. Here we define the range frequency consistency as follows,

$$RangeConsist = (\forall i \in [1, \ldots, k], Count(V_i not \in Range)$$
$$< Max\_OutRange) --- (8)$$

where $RangeConsist$ denotes the range frequency consistency. V_i shows a number of k readings in a time interval and $Max\_OutRange$ denotes the application pre-set maximum number of readings that may be out of the normal range, *Range*, in one period. This consistency can be checked both locally at each sensor and at the sink. Further action are usually needed by the application scientists when this type of consistency is violated.

In some other applications, application scientists may care a lot about the detail of the data changes, thus we define the change frequency consistency to detect whether the changes of the sensor reading are abnormal. The detail of the change frequency consistency is denoted as follows,

$$ChangeConsist = (\forall i \in [1, \ldots, k], Count(|V_{i+1} - V_i|$$
$$> Max\_Diff) < Max\_CHG) - - - (9)$$

where *ChangeConsist* depicts the change frequency consistency. V_i is a set of total k readings in a time interval; Max_Diff is the pre-set maximum difference between two continuous readings when the consistency holds, and Max_CHG means the maximum number of dramatic changes, which is defined as the case that the difference between two continuous readings exceeding Max_Diff, in one interval. With this consistency, we can either prevent the data from changing too dramatically or dynamically change sampling rate to zoom in and observe the details [Sha and Shi, 2006a]. The observation of violation of this consistency may also result in a request of application scientists involvement.

Moreover, if we consider the spatial relationship among sensed data, we also need to satisfy the spatial consistency, which is defined as the consistency of the data among geographically distribution of the data. For example, when we are sampling the temperature, we may have some pre-knowledge of the geographical distribution of the temperature, thus spatial consistency should be checked when the data is collected in the sink. Finally, as we know that there maybe some relationship between parameters, e.g., the speed of vehicles may correlated with the density of the vehicles on the road, we can further explore the data relationship among different parameters by defining consistency models for them.

In summary, we propose a set of basic but powerful consistency models for data quality measurement in WSN from the perspective of temporal, numerical and frequency consistency. These models can be used as metrics to evaluate the quality of collected data both in aggregated format or non-aggregated format. With these proposed basic consistency models, various applications can find their suitable consistency models for their specific data quality requirements by adjusting the parameters in these models or composing the above basic proposed consistency models to form complicated models. For example, the two frequency consistency models can be combined to control the dramatic data changes and the abnormal readings in a time interval. The partial consistency and the two frequency consistency are also composable to set all numerical, temporal and frequency consistency requirements. Furthermore, various applications should make a trade off between the energy efficiency and data consistency based on their energy budget, which remains an open problem in the community.

7.3 APIs for Managing Data Consistency

There is a gap between the lower layer protocols designed to support the consistency goal and the higher layer consistency requirements from applications. It is critical to provide user-friend interfaces for application scientists to take advantage of these models. Our APIs are designed for the purpose of data quality management, and differ from the APIs proposed in [Welsh and Mainland, 2004]. First, our APIs lie at the higher layer (for application scientists) than theirs (for system programmers). Second, the design goals are different too. We believe that our APIs can take advantage of theirs in the real implementation. To manage the consistency, the APIs must have the following functions, *checking the current consistency status* ($CheckStatus$), *setting consistency requirements* for new parameters ($SetReq$), *updating consistency requirements* ($UpdateReq$), and getting support from lower layer protocols, as listed in Table 7.1, where $CommPtn$ denotes the communication pattern to distribute the consistency requirements; $ConMode$ depicts the name of consistency model. Together with V_{req}, T_{req} and other parameters, $ConMode$ also specifies the consistency semantics; and $Set(n)$ depicts the set of destination nodes.

Several protocols and algorithms are needed to support the above proposed APIs. For example, consistency checking algorithms are needed when the $CheckStatus$ API is called. Various algorithms are needed to check consistency in different models. While in $SetReq$ and $UpdateReq$, different protocols are used depending on the size of $Set(n)$. If there is only one node in $Set(n)$, a point-to-point communication pattern is adopted to deliver the consistency requirements. When $Set(n)$ contains all the sensors in the field, broadcast is launched to distribute the requirements. If $Set(n)$ contains nodes located in one area, area multi-cast is used to disperse the requirements. Hence, various routing protocols are needed for different communication patterns.

With the APIs we mentioned just now, we can manage data quality of the collected data based on these models according to application consistency requirements. When the application needs to sample several different parameters. These parameters have different consistency requirements; then we can set different models for these parameters, and check the consistency against these models. For example, in the application of SensorMap [Nath et al., 2007], we can integrate our APIs with the DataHub in the SensorMap architecture. When the data arrives at the DataHub, the consistency model is checked. For each data, suitable consistency model is utilized, so different consistency requirements can be satisfied. For instance, the $k - consistency$ may be set for accident of alarm data, while $ParConsist$ is enough for temperature data.

The process of consistency management consists of three steps. First, the consistency requirements are distributed. Second, the consistency will be checked with support of some consistency-checking algorithm at the sink after an amount of data are collected. Third, if the sink finds that current consistency cannot be satisfied because of the constrained resource, it might release consistency requirements. If the sink find that the quality of current collected data is not satisfactory, it might increase consistency requirements. These update will be distributed to related nodes, who will in turn change their data collecting strategy according to the new consistency requirements.

Next, we give an example of how to use these APIs. In a habitat monitoring application, if an irregular animal movement is observed in some area, a request to monitoring the temperature of that area is issued, i.e., $SetReq(temp, Set(area), T_{req}, Val_{req}, 0, 0, 0, 0, 10\% -$ $LossConsist, area - cast)$, where $temp$ denotes the name of parameter; $Set(area)$ and $area-$ $cast$ show that the consistency requirements will be sent to all the nodes in that $area$ using $area-cast$. T_{req}, Val_{req} and $10\% - LossConsist$ specify the consistency semantics. Four zeros denote no specific frequency consistency requirements. After the application scientist collects some data from the monitoring area, he/she will call $CheckStatus(temp, Set(area), last-$ $one - hour, 10\% - LossConsist)$ to check whether the data received in last one hour satisfy the requirements specified by the consistency mode, $10\% - LossConsistency$. Based on the result of the call, the application scientist makes a decision to tune consistency. For example, if the application scientist thinks that the quality of the data is good enough, he will do nothing to change it; otherwise, he will update the new consistency requirements by calling $UpdateRqe(temp, Set(area), \delta(T)), \delta(Val), 0, 0, 0, 0, 5\% - LossConsist, area - cast)$. Then, when receiving the new update request, the node will update the consistency requirements locally. The whole process forms a close-loop feedback control. In this way, high quality data could be collected in an energy efficient way.

In this dissertation, we propose to use data consistency as a metric to evaluate the quality of collected data in wireless sensor networks. In this chapter, several formal consistency models are defined for different applications. We also propose a set of APIs for the application scientists to efficiently manage data consistency. Next, we need to propose a suit of protocols to provide the system support to achieve the goals in data consistency and energy efficiency, which are specified by various applications and evaluated by those defined models in this and last chapters.

APIs	Function Descriptions
$CheckStatus(P, Set(N), T_{ran}, ConMode)$	Check the consistency of the data from the nodes in $Set(N)$ during the time period T_{ran} according to the data consistency requirements to that data.
$SetReq(P, Set(N), T_{req}, V_{req}, Range, Max\_OutRange,$ $Max\_Diff, Max\_CHG, ConMode, CommPtn)$	Set the temporal or numerical consistency requirements for parameter P to the set of nodes $Set(N)$ using the specified communication pattern to distribute the requirements.
$UpdateReq(P, Set(N), \delta(T), \delta(V), Range, Max\_OutRange,$ $Max\_Diff, Max\_CHG, ConMode, CommPtn)$	Update the temporal or numerical consistency requirements for parameter P at the set of nodes $Set(N)$, and use the specified communication pattern to distribute the updating.

Table 7.1: APIs for data consistency management.

CHAPTER 8

AN ADAPTIVE LAZY PROTOCOL

The consistency models and lifetime models have been described in previous chapters, which can be used as metrics to evaluate the energy efficiency property of system protocols and the quality of the data collected by the wireless sensor system. In this chapter, we propose an adaptive, lazy, energy-efficient protocol to support achieving the goals of energy efficiency and data consistency. In the following sections, the protocol will be described in detail first. Then, the protocol is evaluated in terms of the proposed performance metrics, both in simulation based on TOSSIM simulator and a prototype using 13 MICA2 motes.

8.1 ALEP: An Adaptive, Lazy, Energy-efficient Protocol

In this dissertation, we intend to save energy by estimating the value of the sensing data and adapt the data sampling rate to improve the quality of collected data. This in turn will reduce the number of delivered messages, which is the most significant energy consumption factor in WSNs [Min and Chandrakasan, 2003], Note that the estimated data should satisfy the consistency requirements of applications. The objective of this dissertation is to reduce the number of the delivered message to save energy consumption. In this section, we first introduce the rationale of our design, then give the details of the protocol.

8.1.1 Rationale

As argued in the previous sections and [Sha et al., 2008a], when the data sampling rate is low, there will be more estimated data. The sampling rate affect the data accuracy a lot especially when the estimated value for the data is not so accurate. Estimating data at the sink is used to save energy but it may hurt data accuracy. There are two extremes between data accuracy and energy-efficiency. For energy-efficiency purposes, we can only gather and deliver very small amount of data. Subsequently, the gathered data cannot satisfy the consistency requirements of the application. On the other hand, if we always keep high sampling rate and deliver a lot of messages to get very accurate data, sensors will run out of energy very quickly. Thus, we should make a tradeoff between the energy consumption and the data accuracy. From our observation, we find that the data dynamics varies temporally and spatially. Furthermore, we also find that it is easier to get accurate estimation when the data dynamics is low, however it is difficult to get accurate estimation when the data dynamics is high. Thus, the sampling rate should adapt to the data dynamics in both temporal and spacial ways. When the data dynamics is high, the sampling rate should be raised to improve the data accuracy, otherwise, it should be decreased to reduce the number of delivered energy.

Except adapting the data sampling rate to data dynamics, we can improve the techniques to estimate the next data, so the number of delivered messages can be dramatically reduced using estimated data to replace the sensing data and high data accuracy is kept. Besides, as mentioned in literature [Mainland et al., 2005], sending a message with long length is more energy efficient than sending several messages with short length. Thus, we intend to integrate multiple short messages into one big message.

In summary, our proposed *Alep* protocol consists of three components, *adapting the sampling rate, keeping lazy in transmission based on consistency-guaranteed estimations, and aggregating and using long length packet.* These methods are described in detail in the following subsections.

8.1.2 Two Types of Data Dynamics

The data consistency should also be integrated with the feature of data dynamics in the sensor field. In this dissertation, *data dynamics* means the trend and frequency of data changes. Usually, the data dynamics comes from two dimensions, *temporal data dynamics* and *spacial data dynamics*. In the temporal dimension, data changing frequency varies at different time periods. Figure 8.1 shows the data changing in terms of the time. In the figure, the data changes very fast before time t1 and between time t2 and t3, while it keeps almost stable between time t1 and time t2. Thus, if we keep the constant data sampling rate, the different data consistency will get during different periods with variant data dynamics. On the other hand, from the spacial dimension, the data dynamics diffs from area to area. An example of data changing differing spatially is shown in Figure 8.2. In the figure, the data changes quickly in the right part of the sensor field and slowly in the left part. If we use the same data sampling rate in different locations, we will get different data accuracy, i.e., the collected data may be accurate in the area with low data dynamics, but not accurate for the area with high data dynamics. Furthermore, the temporal data dynamics and spacial data dynamics effect the data consistency at the same time. Thus the data sampling rate should be adapted to the feature of data dynamics from time to time and from area to area. For example, it should sample more data when the data dynamics is high and in the area with high data dynamics, while sample less data when data dynamics is low and in the area with low data dynamics. Moreover, it should use high sampling rate in the area with high data dynamics and use low sampling rate in an area with low data dynamics.

8.1.3 Model for Data Dynamics

Before giving the details of the adaptive protocol, we first model data dynamics. To describe data dynamics, we define a number of windows to observe the data readings. Two parameters, $winSize$ and $winNum$ are defined to model the dynamics of data. $winSize$ denotes the number of readings in one window, e.g., if the $winSize$ is seven, then in one monitoring window, the sensor will obtain seven readings, $winNum$ specifies the number of windows in one observation, e.g., if $winNum$ is four, in one observation there will be four windows. Thus the total number of readings in one observation is $Num_{rd} = winSize*winNum$. Since

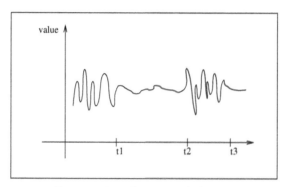

Figure 8.1: Data dynamics with the time.

the data dynamics reflects *the frequency of data changing*, so we first define the frequency of the data changing as the number of data changing in one observation:

$$Num_{chg} = \{Cnt(i)\|r_{i+1} - r_i| > B\&i \in [0 : Num_{rd}]\}$$

where, $Cnt(i)$ is the number of is satisfying the conditions; r_i and r_{i+1} are the ith and $i+1^{th}$ readings separately. $B = C(p)_{bnd}$ is the accuracy bound for this parameter. Based on this definition, we define the data dynamics (DYN) as the average number of changing in one monitoring window.

$$DYN = \frac{Num_{chg}}{Num_{rd}} * winSize$$

From above definition, we can find that the data dynamics is defined based on time period, i.e., inside the window of observation. By adjusting the value of $winSize$ and $winNum$, we can get the data dynamics with various sensitivity. For instance, when we set $winNum$ small, the value of DYN will be calculated with high frequency, i.e., it can be very acute to the data changing. While the value of $winSize$ controls the range of the DYN, e.g., if we set $winSize$ to two, data dynamics can be expressed as above one and

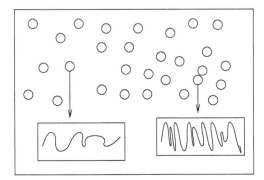

Figure 8.2: Data dynamics in different location.

below one; however, if we set the value of $winSize$ to four, data dynamics can have four levels, below one, one to two, two to three, and above three. Based on data dynamics, it is possible for users to choose suitable data sampling rate to accurately collect data in an energy efficient way, which will be explained in detail in Section 8.1.

For a series of n sensing data, if we get every piece of data, the accuracy is the best by using reading values as estimation values. If we get readings in a half frequency, the accuracy will decrease since we have to estimate half of the data. On the other hand, the energy is saved from sampling and reporting less data. Thus data sampling rate should be decided by making tradeoff between the data accuracy and energy efficiency, which, we argue that, can be achieved by matching data sampling rate to data dynamics. Because data sampling rate should fit data dynamics, we model the behavior of data dynamics here. The dynamics of the data stream can be viewed as two dimensions, temporal and spatial. For example, in the dimension of temporal. Figure 8.1 shows the data reading changing in terms of the time of the same sensor. In the figure, before the time t1 and between time t2 and t3, the data changes very frequently, while it keeps almost stable between time t1 and time t2. An example of data changing differing from different location at the same time is shown in Figure 8.2. In the figure, we can see that the data changes quickly in the right part of the sensor field and it changes very slowly in the left part in the figure. Thus for accurately collecting data in an energy efficient way, the data sample rate should be adapted from time to time and from area to area. For example, it should sample more data when the data dynamics is high, while sample less data when data dynamics is how. Similarly, it should use high sample rate in the area with high data dynamics and use low sample rate in an area with low data dynamics.

8.1.4   Adapting the Sample Rate

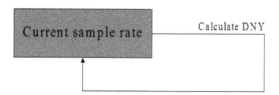

Figure 8.3: Adapting data sampling rate.

We adapt the sampling rate based on the model for data dynamics defined in previous sections. The process of adapting the sampling rate is a process of reinforce learning [Sutton and Barto, 1998] based on the data reading as shown in Figure 8.3, i.e., the data is sampled based on the current sampling rate; then the data dynamics is calculated; finally, the data sampling data is adjusted to fit the data dynamics. In the model, data dynamics reflects the average number of changes in one monitoring window. Thus based on the value of DYN, we can define the adaption of the sampling rate as

$$R_{smp} = \begin{cases} \lceil \frac{DYN - Ave_{chg}}{Df_{bnd}} \rceil * R_{cr}, & DYN > Ave_{chg} \\\\ \frac{Ave_{chg} - DYN}{Df_{bnd}} * R_{cr}, & DYN \leq Ave_{chg} \end{cases}$$

where, R_{smp} is the adapted sampling rate; R_{cr} is the current sampling rate. Ave_{chg} is the normal average changes happen in one window size; and Df_{bnd} bounds maximum difference between the observed value of data dynamics and the normal average changes, i.e., if DYN is larger than Ave_{chg} and the difference exceeds the bound, the sampling rate should increase; when DYN is much smaller than Ave_{chg}, the sample rate should be decreased.

Based on this formula and the figure, the sampling rate learns from the previous data dynamics, and uses the most recent data dynamics to estimate the future data dynamics. The data history is limited by the number of windows and the window size in one observation. The result of adaption divides the sample rate to several levels based on the value of bound, and find a suitable level for the future data collection. By adjusting the length of history based on the window size and the number of windows, we can adjust the frequency of sample rate changing and the acuteness of the changing of the environment, e.g., we can change the sampling rate very quickly by setting small value to the number of windows in observations.

8.1.5 Keeping Lazy in Transmission

One way to reduce the number of delivered messages is to keep lazy in transmission, i.e., only sending the messages that are necessary to be sent. We think that if the receiver can estimate an accurate enough value for the current reading, the message need not to be sent, i.e., if the data consistency requirement can be hold, the messages is not necessary to be sent.

In this protocol, every sensor caches the last reading for every parameter for all potential senders that may deliver message to it, and it uses the cached values as the estimation of the current reading. To check the data consistency for this piece of data, the sensor will use the current reading as the real value and the cached value as the estimated value. If the difference between the current reading and the cached value is within the consistency bound, the sender will not send this piece of data, i.e., keeping lazy. For example, in an application which monitors the temperature of a sensor field, when a sensor gets a reading of value 3.7, and the cached last reading is 3.5 which is within the consistency bound of 0.3. So the new reading is not necessary to be sent. When the current data reading is absent, the sensor assumes the value is unchanged so that it keeps silent.

In the case of the aggregated data, every receiver caches a copy of the latest aggregated value calculated from senders. After it applies the aggregation function, it will compare the new calculated value with the cached value. If the difference between them is within the consistency bound, the sender will keep silent. For the aggregated data, the receiver has to wait for the new reading from all the senders for a period of time. If there are still data absent from some senders, the receiver will use the cached data to substitute the current reading and calculate the aggregated value.

8.1.6 Aggregating and Delaying Delivery

Another aspect of the lazy approach is to integrate several pieces of data into one message to reduce the number of message. However, we still need to keep the timeliness for the data. In our application, the expected time to deliver the message to the sink can be estimated based on the number of hops to the sink. For example, if we assume it takes T_{dev} to transmit one message from the child to the parent, then we can estimate the time it takes from current sensor to deliver a message to sink is $T_{dev} \times H_{js}$, where H_{js} is the number of hops from the current sensor to the sink. Then the time bound for the data is the sum of the current estimated time plus one time slot, which denotes the time between two reporting points according to the TDMA schedule. Using the estimated time to the sink, we can easily check the temporal consistency requirement for the parameter.

8.1.7 Discussions

Note that the proposed protocol is a general protocol for sensor networks, we still have several assumptions. First, the data readings from sensors are accurate, i.e., here we do not consider reading errors. Second, synchronization between sensors are kept by using some mechanisms, e.g., RBS [Elson et al., 2002]. Third, sensors in the sensor field are static, which is true in most monitoring-based applications. Finally, the sensors in the sensor field are homogeneous, i.e., each sensor has the same physical capacity.

Although our protocol is proposed for a tree-based sensor network, the basic idea of adapting the data sample rate, keeping silent during transmission and merging transmissions can be applied in any protocol to achieve both the data accuracy and energy efficiency. In a tree-based sensor network, this protocol can be implemented by an ideal TDMA schedule.

From the model for the problem definition, we can see that the optimization at a single sensor can guarantee the system level optimization. However, the system level optimization doesn't necessarily require the optimization at single sensor. The consistency requirement at a single sensor is more rigorous than the consistency requirement for the whole sensor network. So, we may loosen the consistency requirement for individual sensor a little, and the consistency requirement for the whole sensor network will still be hold at a very high probability.

In our design, we use the last reading to estimate the previous reading which is consistent when the transmission is reliable. However, it only reduced the messages having data within the consistency bound, when the current reading of the data is in the endurance bound of the real reading. If other techniques that can accurately estimate the value of the data out of the range of the consistency bound, it will further reduce the number of delivered messages, which will be our future work.

8.2   Performance Evaluation: Simulation

To evaluate the performance of the proposed protocol, we have implemented the protocol in TinyOS using the TOSSIM [Levis et al., 2003] environment and compared with two other protocols, *Simple* which is a TDMA-based data collection protocol and *Lazy* which only has the lazy feature of the proposed protocol. In the rest of this section, we will describe the simulation setup and the performance metrics first, followed by the performance evaluation in simulation environment. The results of a prototype implementation and evaluation is reported in the following section.

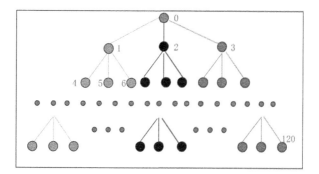

Figure 8.4: A tree-structured sensor network used in the simulation and prototype. The numbers next to each node is node ID or Mote ID.

8.2.1 Simulation Setup and Evaluation Metrics

In our simulation, 121 nodes are connected forming a four layer complete tree, as shown in Figure 8.4, where all the internal nodes have three children and the root acts as the sink. The sensors periodically collect data from its children and report the readings to its parent based on a TDMA schedule. Besides, the sensors may have the ability of aggregation. The data consistency requirements at each sensor are preloaded from the sink by broadcast. The whole circle area is divided into three sub-areas, which is covered by a sub-tree. Each sensor node acts as a multiple functional sensor, which can sample three parameters: Temperature as Temp, Pressure as Press, and Rain-index as Humid. To evaluate the proposed protocol in different data dynamics environments, we intentionally make these three parameters have different dynamic characteristics. For example, the reading always changes faster for Temp, relatively stable for Press, while medium for Humid. To simulate spatial data dynamics, we intentionally separate the whole area into three sub-areas with different data dynamics as shown in different colors (gray levels in B/W print out) in the figure. The reading changes faster in the left subtree area, relatively stable in the right subtree area, and medium in the middle subtree area.

For the perspective of spatial, the reading of each parameter will differ based on the location. Basically, in our simulation, the whole area is separated to three sub-areas as shown in Figure 8.4, where the data changing in the area covered by the left subtree is always fast, while slow in the area covered by the right subtree, and normal in the area covered by the middle subtree. Three algorithms will be simulated in our experiment. First, the method without considering both lazy and adaptive approaches will be simulated, denoted as *Simple*. Second, the method with lazy approach but without adaption is simulated, denoted as *Lazy*. Finally, the method considering both lazy and adaptive approaches is simulated, denoted as *Alep*, where we try to keep $\alpha-$Consistency where $\alpha = 20$. The consistency is checked at each sensor as well as the base station.

The goal of the *Alep* protocol is to save energy by reducing the number of delivered messages while satisfying the data consistency requirements. Thus, we use three metrics to evaluate our approach. To measure the energy efficient property, we count the total number of delivered messages and the dropped voltage at each sensor (in prototype evaluation), and to examine the tradeoff between the energy efficiency and data consistency. Thus, *Alep* will be examined in three ways: *Does this protocol reduce the number of the messages and extend the lifetime of WSN? Does this protocol improve the accuracy of data? What is the tradeoff between the number of delivered messages and the data accuracy?* To answer the question of the effect of reduced messages to data consistency, we propose a new performance metric called *data inconsistency factor (DIF)*, which is defined as the total variance between the gathered data in the sink and real data, i.e., $V = \sum_1^n (d_{rcv} - d_{fld})^2$, where, V is the value of variance; d_{rcv} and d_{fld} are the reading value received at the sink and the real value sampled at the data field respectively. The more accurate the data, the smaller the variance.

To examine the tradeoff between the energy consumption and the data accuracy, we adjust the value of the temporal consistency bound and the numerical consistency bound, which are two parameters in the *Alep* protocol. By adjusting these parameters, we can get different simulation results in terms of energy consumption and data accuracy. we will adjust the parameters in the *Alep* protocol to show the effect of these parameters to the energy consumption and data accuracy. These parameters includes the bound of both time accuracy and value accuracy.

8.2.2 Number of Delivered Messages

Usually collecting more data is a way to improve the data accuracy; however, by adapting the sampling rate to fit the feature of data dynamics and keeping lazy when data is in the range of consistency, data accuracy can be improved without significantly increase the number of delivered messages. Moreover, in some cases when the data dynamics is low, the data consistency can be kept even by delivering less number of messages. In this section, we show the number of messages delivered at each sensor using different approaches.

Figure 8.5 lists the number of delivered messages at each sensor without and with aggregation respectively. The x-axis is the ID of each sensor, and the y-axis denotes the number of delivered messages. Note that the y-axis of Figure 8.5 (a) and 8.5 (b) are at different scales. As a matter of fact, the number of delivered messages for all approaches reduces significantly when aggregation is used. From the two figures, we can see that *Simple* generally delivers the most number of messages and *Lazy* transfers almost the least number of messages in both cases of with and without data aggregation. That is because *Lazy* filters a lot of unnecessary messages.

These three approaches have totally different performance in terms of the number of messages delivered. In the case of without data aggregation shown in Figure 8.5 (a), the sensors are classified to four types based on the layer in the tree using *Simple*, i.e., sensors in the same layer using *Simple* delivers the same number of messages. However, using *Alep* and *Lazy*, the sensors transmit different number of messages because of the variant data dynamics in the different areas. For example, among sensors located at layer 3, sensors with ID between 13 and 21 transfer 140 messages because the high data dynamics of the monitoring area, while the sensors with ID between 31 and 39 only deliver 41 messages because the low data dynamics of the monitoring area, which is less than $\frac{1}{3}$ of that in the high dynamics area. The similar results exist in the case with data aggregation in Figure 8.5 (b), where all the sensors deliver the same number of messages using *Simple*, while the sensors using *Alep* and *Lazy* located at different areas transmit different number of messages, i.e., the sensors located at high dynamics area deliver 57 messages but the sensors located at low dynamics area only send 9 messages, which denotes that *Alep* does adapt the data sampling rate to the dynamics of the data

Comparing with *Lazy*, we observe that the sensors using *Alep* send more number of messages than using *Lazy* at the area with high data dynamics (e.g., node 13 – 21) but send less number of messages than that of using *Lazy* at the area with low data dynamics (e.g., node 31 – 39). This is because the sampling rate is increased considerably in the area with high data dynamics and decreased a lot in the area with low data dynamics. From above analysis, we conclude that *Lazy* can always reduce the number of delivered messages, and *Alep* usually does not increase the number of delivered messages and reduce the number of delivered messages a lot when the data dynamics is low.

To take full advantage of adapting sample rate, we need an intelligent adaptation scheme, which is our future work. In the following sections, we will see that reducing of the number of messages does not necessary degrade the consistency of the collected data.

8.2.3   Data Inconsistency Factor

From above sections, we can see that *Lazy* and *Alep* can largely reduce the number of delivered messages. However, delivering less message means that there are more data estimated at the sink, which may result in the degradation of the data consistency. In this subsection,

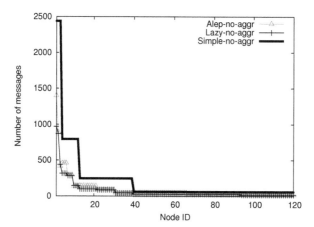

Figure 8.5: Number of delivered messages without aggregation.

we examine the effect of unsent messages to the data accuracy. We use data inconsistency factor as the metric to measure the effect.

We use the data inconsistency factor to evaluate the quality of the collected data. Figure 8.7(a) reports the relationship between the data inconsistency factor and different monitoring parameters with variant data dynamics. In the figure, the x-axis is different data types with variant data dynamics and the y-axis represents the calculated data inconsistency factor of the collected data. Three types of parameters with different data dynamics are monitored, among which Temp has relatively higher data dynamics than Humid and Press, while Press has relatively lower data dynamics. Furthermore, for each parameter, data dynamics also varies according to different areas, i.e., each parameter has three types of data dynamics, high, high first then low denoted as mix, and low. Thus, there are totally nine sets of data with variant data dynamics.

In the figure, we note that when the data dynamics is higher, the value of data inconsistency factor is larger, e.g., the Temp high has a larger data inconsistency factor than

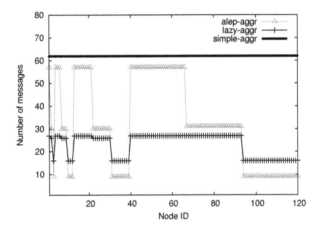

Figure 8.6: Number of delivered messages with aggregation.

Temp mix and Temp low, and Temp high also has a larger data inconsistency factor than Humid high and Press high. The reason is that when the data dynamics is high, it is more difficult for the sink to estimate the correct data. From the figure, we also find that *Alep* has much smaller data inconsistency factor than that of *Simple* and *Lazy* when the data dynamics is high, while it has larger data inconsistency factor than that of *Simple* and has the same data inconsistency factor as *Lazy* when the data dynamics is low. This result shows that *Alep* indeed makes the sampling rate fit the feature of data dynamics, i.e., when the data dynamics is high, it will use higher sampling rate to gather more data so that to make the variance small. Otherwise, it will sample less data to save energy.

Furthermore, the data inconsistency factor increases very fast with the increasing of data dynamics using *Simple* and *Lazy*, but increases slowly using *Alep*. As a result, *Simple* and *Lazy* may not collect enough accurate data when the data dynamics is high, i.e., the data inconsistency factor exceeds the data consistency requirements of the application. However, *Alep* can keep the data inconsistency factor low by adapting the data sampling rate to data

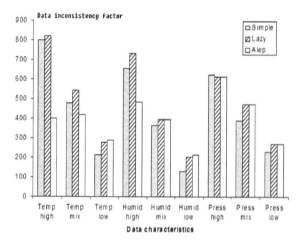

Figure 8.7: Data inconsistency factor.

dynamics. We should also notice that *Alep* improves the data accuracy meanwhile somehow reduces the number of delivered messages as shown in Section 8.2.2.

Comparing *Lazy* with *Simple* in terms of the accuracy of the collected data, *Lazy* has very close value of data variance as *Simple*, however, in Section 8.2.2 we know that *Lazy* delivered less messages than *Simple*, which means that the dropped messages are not necessary to be transferred to the sink. Thus, we conclude that lazy delivering can reduce the number of delivered messages, while the approach of adapting the data sampling rate to data dynamics can significantly improve the data accuracy. It is good to integrate those two approaches to collect accurate data in an energy-efficient way.

8.2.4   Tradeoff between Energy Efficiency and Data Consistency

We have already seen that *Lazy* and *Alep* can largely reduce the number of delivered messages so that they have potential to save energy and extend the lifetime of WSN, and *Alep* can

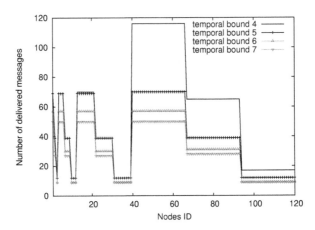

Figure 8.8: Number of messages with variant temporal bound using *Alep*.

sufficiently improve the data consistency. Now we are in position to examine the effect of two key factors related with *Alep*: *the temporal consistency bound* and *the numerical consistency bound*.

The performance of Alep is largely decided by the two key factors, *the temporal consistency bound* and *the numerical consistency bound*. Here we study the effect of these two factors.

First let us consider the effect of the temporal consistency bound to the energy efficiency and data numerical consistency. If we release the temporal consistency of data, the same set of data will be delivered to the sink regardless of different arrival times. Thus changing the temporal consistency bound will not affect the data inconsistency factor of the collected data. However, releasing the temporal consistency bound does affect the number of delivered messages. Figure 8.7(b) displays the relationship between the number of delivered messages and the different temporal consistency bounds ranging from 4 units to 7 units, which is the maximum time to transfer a message to the sink assuming each hop taking one unit time. In the figure, the x-axis is the ID of the sensors and the y-axis is the number of delivered messages. From the figure, we can see that the increasing of the bound of temporal consistency results in the decreasing of the number of total delivered messages. When the temporal consistency bound is tight as 4, some sensors deliver more than 110 pieces of messages because data combination is not possible. While the temporal consistency bound is raised to 7, sensors deliver only about 50 pieces of messages. Thus, releasing the bound of temporal consistency can reduce the number of delivered messages. However, based on simulation data, the energy consumption almost keeps the same (overlapped in the figure) with the releasing of the bound of the temporal consistency as show in Figure 8.8. This is because the same reason of idle listening. Subsequently, this will reduce the energy consumption. In this case a well designed schedule is needed to save energy from idle listening. This problem may be solved automatically in the new version of Motes, such as TelosB [Polastre et al., 2005], which can automatically transfer to the sleeping state.

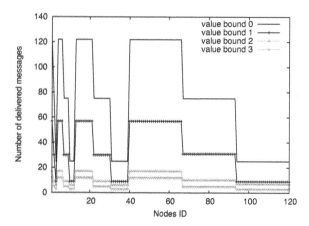

Figure 8.9: Number of messages with different numerical consistency bound.

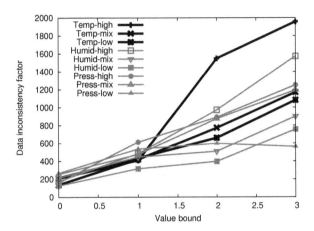

Figure 8.10: Data inconsistency factor with variant value bound using *Alep.*

Having seen that releasing the temporal consistency bound can reduce the number of the delivered messages, next, we examine the effect of the numerical consistency bound to the number of delivered messages and the data inconsistency factor. We also examine the effect of the numerical consistency bound. Figure 8.9 shows the number of delivered messages with the relation to the variant value constraints. In the figure, the x-axis is the ID of the sensors and the y-axis shows the number of delivered messages. From the figure, we can see that when the numerical consistency bound is enlarged, the number of the delivered messages is decreased very fast. Next, we examine the changing of data inconsistency factor with the changing of the numerical consistency bound. Figure 8.10 shows the relationship between the data inconsistency factor and the value of the data consistency bound. The x-axis is the different value bounds and the y-axis depicts the value of the data inconsistency factor. In the figure, when the data consistency bound is released, the data inconsistency factor increases very quickly, especially when the data dynamics is high. Thus we argue that there is a tradeoff between the data consistency and the energy efficiency. Releasing the data consistency bound results in both energy efficiency and larger data inconsistency factor, so the application should decide the data consistency bound based on its specific data consistency requirements. If the application cares little to the data consistency, it may raise the bound, otherwise, it has to use a tighter bound.

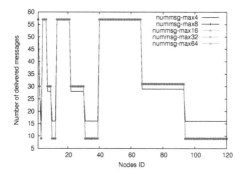

Figure 8.11: Number of delivered messages with variant maximum data sampling rate.

Besides the temporal consistency bound and data consistency bound will effect the variance and the energy efficiency. Another factor, the maximum data sampling rate will also affect these two performance. Here we check the effect of the maximum data sampling rate, which is the maximum data sampling rate *Alep* may take. Figure 8.11 denotes the number of delivered message with different variant maximum data sampling rate. In the figure, the x-axis is the ID of the sensors and the y-axis shows the number of delivered messages. We can find that the maximum sampling rate will not affect the number of delivered messages a lot. Only when the sample rate change to very small, it will sample more data when the data dynamics is low.

8.3 Performance Evaluation: Prototype

Although the simulation results have shown the advantage of Alep and Lazy, we further evaluate them in a prototype implementation, where energy consumption is measured more directly and accurately. In the prototype, 13 MICA2 Motes form a tree similar to that in the simulation except that it has only two layers, the root of the tree is connected to a desktop. Other configurations of each sensor is the same as that in the simulation. Each Mote samples three parameters with different data dynamics and the whole tree is divided into three subareas with various data dynamics. For comparison purposes, three algorithms are implemented, including Alep, Lazy and TinyDB, a simplified version of TinyDB [Madden et al., 2005] without data aggregation.

In the implementation, we find that the program developed for TOSSIM could not be executed at MICA2 Motes directly because TOSSIM does not enforce the same strict memory constraints as that in MICA2 Motes. However, to compare these three protocols, we need to feed them enough data, either synthesized or real traces, to show the differences. This is a challenge in the prototype implementation because we cannot use the data sampled from the sensor board directly. As a substitution, trace based approach and a data generator are considered. However, we argue that the trace based data feeding does work in sensors smoothly. First, it is impossible to hold a long trace in the program flash memory of tiny sensors, which have only 128k bytes in total. Second, there are also some disadvantages to store the trace file at the measurement flash of MICA2, which has 512k bytes in total. For example, as denoted in [Shnayder et al., 2004], the current for reading MICA2 sensor board is 0.7mA, while the current for reading and writing EEPROM is 6.2mA and 18.4mA separately. These overheads are comparable with that for receiving and sending messages. Thus, a lot of energy will be consumed by reading data from measurement flash so that the energy consumption of the protocol cannot be accurately evaluated. Furthermore, the total possible access time of the measurement flash is limited [Shnayder et al., 2004]. So, we decide to design a data generator for evaluation purposes. In our prototype we decide to design a data generator to produce the same set of data as the sampling feedings for all protocols. The detail of the data generator is described as next subsection.

8.3.1 SDGen: Sensor Data Generator

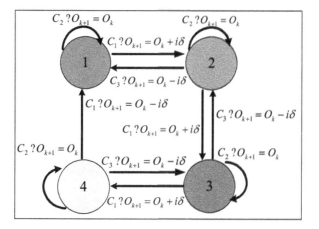

Figure 8.12: State transition diagram of sensor data generator.

We intend to design a general sensor data generator, SDGen, which can be used by other researchers for evaluating their protocols and algorithms as well. SDGen is a simple data generator based on the a finite state machine as shown in Figure 8.12 (a), which intends to generator the same series of random readings. In the figure, the four circles denote the four states. SDGen remembers the latest output as O_k and a state as one of the four circles in the figure. To generate a new sampling, it first generates a random number. Based on the value of the random number, the conditions, C_1, C_2, and C_3, are checked and the state of the machine is transferred. The action after the satisfied condition is executed and an new output O_{k+1} is produced based on O_k. For example, if C_1 is in the range of [0.0, 0.9], and the random number generated is 0.6, then $O_{k+1} = O_k + \delta$ is the output and the machine transfer state from 1 to 2; if C_2 is in the range of $(0.9, 1.0]$, and the random number generated is 0.95, then the output is $O_{k+1} = O_k$ and the machine state is not changed.

Our SDGen has several advantages. First, it is very easy to implement and it only needs to remember the last state and the last output, which reflects the fact that the next sampling is usually closely related with the latest previous reading. Of course, we can make it more complicated by remembering more previous readings. Second, the same set of random data can be generated if we fix the value of the random seed, which satisfies our requirements of feeding the same set of data to all protocols. Actually, in our implementation, we find that it is important to keep the order of the generated random number, because we need to gradually generate samplings for three parameters. We take the following strategy in our implementation: all samplings are read from a short array, which stores a set of latest generated data and the order of the generation is controlled by the length of the array. Third, SDGen can generate sampling series with different data dynamics by adjusting the parameters, including the value of i and the condition C_i. For instance, if we make C_2 to be in $[0.0, 0.8]$, which means that any two continuous samplings are the same with a probability of 80%, dynamics of the generated data will be very low, while if we make C_2 to be in $[0.0, 0.1]$, dynamics of the generated data will be much higher.

8.3.2   Comparison Number of Delivered Messages

As argued in [Min and Chandrakasan, 2003], the number of delivered message dominates the energy consumption in WSN applications. Thus we first compare the number of delivered messages using these three protocols. As hinted in [Sha and Shi, 2004, Sha and Shi, 2005], the lifetime of a WSN is decided by a set of communication intensive sensors, which are layer one nodes in a tree-based structure. Thus, we compare the number of delivered messages of these layer one nodes only.

Figure 8.12(b) shows the results of number of delivered messages. In the figure, the x-axis is the Mote ID, which also depicts the area with specific data dynamic feature and

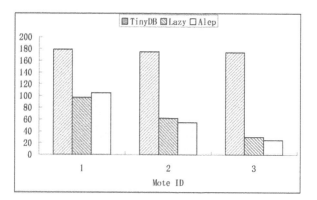

Figure 8.13: Comparison of number of messages delivered by Layer one Motes.

the y-axis is the number of received messages at the sink from the corresponding Mote. We find that in all cases, both Alep and Lazy send much less number of messages than TinyDB. For example, when data dynamics is high, TinyDB sends about 170 messages, while Alep and Lazy send only about 100 messages. When data dynamics is low as for Mote 3, TinyDB still sends about 170 messages, but both Alep and Lazy send less than 30 messages. With the gradual decreasing of data dynamics from Mote 1 to Mote 3, TinyDB sends almost the same number of messages, while both Lazy and Alep send less messages. Compared with Lazy, Alep sends comparable number of messages in all cases; however, it sends a little more messages when data dynamics is high, and sends a little less messages when data dynamics is low. This denotes that Alep does adapt the data sampling rate to match data dynamics.

8.3.3 Comparison of Energy Consumption

We use the voltage drop to show the energy consumption in these three protocols as shown in Figure 8.14, where the x-axis is the Mote ID and the y-axis is the value of the initial and final voltage read from every node using different protocols. The initial voltage is measured

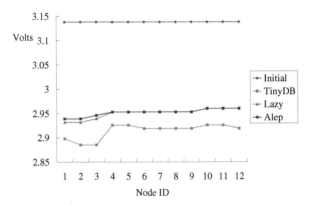

Figure 8.14: Comparison of voltage dropping of different protocols.

before the protocol is executed and the final voltage is measured after 50,000 samples are collected. We use the same brand new AA Alkaline batteries in all the experiments to get the same initial voltage, 3.138 volts, and we find that different protocols result in different final voltages. TinyDB consumes much more energy than the other two. For example, the voltage drops 0.24 volts at layer one nodes and 0.22 volts at layer two nodes in TinyDB, while it drops only 0.20 and 0.18 volts correspondingly in both Lazy and Alep. As expected, layer one nodes consume more energy than layer two nodes because they need to forward messages for the latter. Motes in the same layer consume similar energy. The areas with different data dynamics consume almost the same amount of energy, because the number of messages does not differ too much; however, we still see that the area with lowest data dynamic does consume less energy than the Motes in other areas. Lazy and Alep have very close energy consumption, however lazy consumes less than Alep, especially for the layer one nodes. In summary, we find that the energy consumption mostly matches the result of the number of delivered messages. Alep and Lazy are really energy efficient, and they can extend the lifetime of WSNs considerably.

8.3.4   Comparison of Samplings

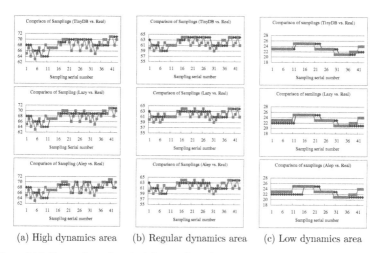

(a) High dynamics area   (b) Regular dynamics area   (c) Low dynamics area

Figure 8.15: Comparison of received data at the sink with real data. Note that the SQUARE shape represents the values of real data and the DIAMOND shape shows the value of collected data.

We have shown the energy efficiency property of Alep. Next we want to show that Alep can also improve data quality. The most intuitive way to show the quality of collected data is to compare them with the corresponding data generated by SDGen, acting as real data. In this experiment, we sample every 4 readings from all data generated by SDGen in TinyDB and Lazy, and the sampling rate is changing in Alep because of adaptation. Other three readings will be estimated as the same as the sampled one.

In Figure 8.15, three types of data with variant data dynamics are shown with the x-axis denoting the sampling serial number and y-axis depicting the value of the corresponding sampling. The figures should be read column by column. When data dynamic is high as high as with a probability of 90% to change, as show in Figure 8.15(a), all three protocols perform similar at the beginning, but Alep catches the trend of the sampling better than the other two protocols because it adapts to a high sampling rate. Thus more details are observed, for example, between sampling serial number 21 and 40, Alep catches the dynamics of the data very well, while both Lazy and TinyDB miss the details. Furthermore, Lazy filters more details than TinyDB, but we argue that the degrading of data quality will not be significant because it is bounded by the consistency endurance range of Lazy. In Figure 8.15(b), data dynamics is not as high as that in (a) by setting the data changing probability to be 60% in SDGen. All three protocols collect very similar data, which captures the dynamics of real data. In this case, although Alep seems not perform better than TinyDB, it delivers much less messages than TinyDB. Moreover, we can adjust the parameter in Alep so that the adaptation can be sharper, e.g., we can set small window size and decrease the threshold of adaptation. Alep can capture more details of the data changing. The data collected from the low dynamic area is shown in Figure 8.15(c), where we observe that all three protocols can capture the trends of the sampling very well. However, Alep catches the trend a little later than TinyDB and Lazy, because Alep decreases the sampling rate when the data dynamics is low. As a result, some immediate changes are ignored and postponed to next time when the data is sampled. This will not be a problem if we set the minimum data sampling rate close to the regular data sampling rate in TinyDB. Comparing Lazy with TinyDB, we find that they are mostly comparable, especially when the data dynamic is not very high, so we can save a lot of energy by using Lazy.

8.3.5   Comparison of Data Inconsistency Factor

We have seen the advantage of the Alep protocol directly and intuitively from the sampled data readings. We also want to compare these protocols quantitatively in terms of data inconsistency factor as shown in Figure 8.16, which the results for three layer one nodes

	Mote 1			Mote 2			Mote 3		
	TinyDB	Lazy	Alep	TinyDB	Lazy	Alep	TinyDB	Lazy	Alep
High dynamics	772	748	678	495	783	409	517	515	629
Mixed dynamics	558	498	610	205	309	341	532	658	624
Low dynamics	116	346	354	87	187	229	136	234	546

Figure 8.16: Comparison of data inconsistency factor.

only. From the figure, we can see that the result is close to that in simulation. When data dynamics is high, Alep can reduce data inconsistency factor, but while data dynamics is low, Alep increases data inconsistency factor a little. However, we argue that we can control the increasing of data inconsistency factor by limiting the minimum sampling rate during adaptation, from which we still can get the advantage of reducing a large amount of messages. Furthermore, data inconsistency factor will be much less if we think the data within tolerance range as consistent in the calculation of data inconsistency factor. Compared with Lazy, Alep has larger data inconsistency factor when the data dynamics is very low, because Alep neglects a lot of details by decreasing the sampling rate.

8.3.6 Discussions

In summary, from both simulation and prototype evaluation, we find that Alep can improve data quality when the data dynamic is high and reduce the number of delivered messages a lot in almost all cases. We also find that the minimum sampling rate is very important to control the quality of collected data, which decreases very fast if the sampling rate is too low to get the details of data changing. Releasing the temporal consistency requirements is helpful to reduce the number of delivered messages so that to save energy, as well as to decrease the probability of communication collision and increase the possibility of data aggregation, so we expect less package loss rate in Alep and Lazy than that in TinyDB. However, it may also make the messages piled and exceed the limitation of the total memory of MICA2. Thus, we should set a low bound for temporal consistency requirements.

We also realized that trace-based approaches only work in the simulation but fail in the prototype test due to the memory constraints. Compared with TinyDB, Lazy and Alep may have more strict requirements on the correct delivery of the messages because the effect of data loss will be more severer than that in TinyDB. Retransmission strategy may be applied here. Furthermore, TinyDB's optimization when message queue is full can also be applied in Alep and Lazy, however, we argue that the probability that message queue is full in Lazy and Alep is much less than that in TinyDB because the number of total delivered messages is significantly reduced. Finally, the performance of Alep can be improved by tuning the parameters, which deserves further investigation.

CHAPTER 9

D4: DECEPTIVE DATA DETECTION IN DYNAMIC SENSING SYSTEMS

We envision that the successful of wireless sensing systems is determined by the quality of the collected data, and the quality of the collected data is mainly affected by the deceptive data, which usually comes from two sources, wrong readings resulted from inaccurate sensing components and unreliable wireless communication, and false data inserted by malicious attackers. Thus the major concern to improve the data quality is to detect and filter deceptive data. The problem of how to improving data quality have been studied in several previous efforts, such as security approaches and reputation based approaches, but we argue that those approaches are necessary to improve the quality of the collected data, while they are not sufficient to attack the problem of data quality management, especially in a highly distributed highly dynamic environment such as vehicular networks.

In this dissertation, we intend to propose a general framework to detect the deceptive data from the view point of data itself. Basically, we try to filter two types of deceptive data, *redundant data* and *false data*. In our framework, those two types of deceptive data are treated differently. Quality-assured aggregation and compression (Section 9.2) is used to detect and filter redundant data, while role-differentiated cooperative deceptive data detection and filtering (Section 9.3) and self-learning model-based deceptive data detection and filtering (Section 9.2) are utilized to filter false data. Finally, when both types of deceptive data are checked and recognized after the data are delivered to a central server, a spatial-temporal data quality checking can be performed to further detect and filter the remaining deceptive data. The novelty of our approach exists in two-fold. First, to the best of our knowledge, we are the first to propose a general framework in deceptive data detection and filtering, and we are the first to propose role-differentiated deceptive data detection and filtering.

Among the several mechanism to detect and filter deceptive data, in this dissertation, we focus on the role-differentiated deceptive data detection and filtering mechanism, and we apply this mechanism in the scenario of a dynamic wireless sensing system called vehicular network, which is composed by a large amount of vehicles with sensing, computation and communication functions. The basic idea of role-differentiated deceptive data detection and filtering is described as follows. First, the vehicles in the system are classified into several groups, road side unit (RSU), public vehicles such as police cars, school cars, and normal vehicles. Each group plays different roles and has different impacts in checking the reported data. For example, in such a system, both RSU and public vehicles have higher trust level than normal vehicles if an event is reported or confirmed by them, and by take advantage of some physical parameters, a vehicle credits most to its owe observation to the event. The reported event will be regarded as false report when there is no sufficiency information about the report is confirmed, thus it will be filtered by the system. In this way, we detect and filter false reports and deliver the legitimated reports as fast as possible.

The rest of this section is organized as follows. Section 9.1 shows the necessary to detect and filter deceptive data in new ways, which is followed by the design of a general framework to detect and filter deceptive data in Section 9.2. Among the several mechanisms proposed in the framework, we give detailed description of the role-differentiated cooperative deceptive data detection and filtering mechanism in Section 9.3. Section 9.4 describes an application of the proposed mechanism in vehicular networks.

9.1 Necessity of Deceptive Data Detection and Filtering

A lot of wireless sensing system applications have been launched in last several years, including habitat monitoring, environmental monitoring, acoustic detection, and so on. Furthermore, vehicular networks and healthcare personal area sensing systems become widely applied recently. Among those applications, The major function of the sensing networks is to collect meaningful and accurate data. In those wireless sensing system applications, two types of data are mostly interested, sensed readings of monitored parameters such as temperature in the room, and detected events like an appearance of the enemy. According to the feature of actions to those collected data, we classify them into two main categories, emergency data and regular data. When the emergency data is collected and reported, a followup action should be taken to take care of the detected event in a prompt way. For example, in a high performance computing system, if it is detected that the temperature of one node is too high, we may have to move workload from this node to others as soon as possible. While regular data, which usually has no restrict timeliness requirements, can be all the sensor readings about the monitoring parameters except those emergency readings. Although those two types of data have different features, they share some common requirements in terms of deceptive data detection and filtering.

9.1.1 Deceptive Data Definition

In this dissertation, we classify the deceptive data into two categories, *redundant data and false data*, based on the affect of the deceptive data to the performance of the sensing systems. The redundant data is defined as the data that shares the exactly same or very similar information with data reported in previous time slots or by other nearby sensors. Most redundant data should be detected and filtered because of following reasons. First, as specified in [Sha and Shi, 2008], in most cases removing redundant data will not degrade the quality of collected sensing data. Moreover, a lot of redundant data will ruin the performance of sensing systems. For example, redundant data will increase the communication, storage and computation overhead when they are transferred from the sensors to the sink. In this case, a lot of limited resources are wasted. Even worse, if the network traffic and storage are occupied by these deceptive data, a lot of meaningful data have to be delayed in transmission or even dropped because of running out of storage, thus the quality of the collected data will be largely degraded. Therefore, it is necessary to filter the redundant data and save resources to deliver more important data. On the other hand, these redundant data provide us replicated information, which is very helpful in building a fault-tolerant sensing system. Thus, there is a tradeoff on detection and filtering these redundant data. We argue that we will adapt our redundant data detecting and filtering protocol to the requirements of sensing system applications, which will decide the redundance level of the collected data that is needed for fault-tolerant purpose.

Another type of deceptive data are false data, which may result from the following several sources. First, false data may be caused by the malfunction of the sensor board so that wrong sensor readings are generated. Second, due to the unreliable wireless communication and limited resources in wireless sensing networks, data may be lost or changed in transmission because of collision. Finally, because of the limited capability in providing high level security and adverse deployment environment of the sensors, sensors may be captured and compromised easier than regular computers. As a result, these compromised sensors can insert false information to raise malicious attacks. False data will degrade data quality in two ways. On one hand, false data is similar to redundant data in that it occupies limited resources and increases the probability of dropping meaningful data. On the other hand, things can be even worse because false data can bring diaster to the sensing system applications. For instance, false data can directly affect the quality of the collected data in that it increases the data inconsistency factor defined in [Sha and Shi, 2008]. Furthermore, false data can cause wireless sensing system in malfunction. For example, wrong traffic information can direct drivers to take some paths with high traffic but avoid some paths with low traffic. In some worse cases, the false data like an emergency brake notification in vehicular networks may result in more severe consequences such as collisions. Thus, it is critical to detect and filter false data as much as possible to avoid the disasters it may bring.

In this dissertation, we are more interested in detecting deceptive data in highly dynamic systems, because in such systems, the sensing nodes behavior changes fast resulted from a variety of reasons, necessitating a quick-response and filtering mechanism.

9.1.2 Insufficiency of Previous Approaches

Security technologies using traditional cryptography mechanisms, such as encryption for confidentiality, hashing digest for message integrity, are employed. However, we argue that these technologies are necessary in detecting and filtering deceptive data but they are not enough in detecting deceptive data, because of mostly they are trying to prevent attackers. Follows are several reasons. First, most of those security based approaches try to prevent attackers, but they rarely check the data themselves. To be specific, they try to validate the legitimation of the reporting nodes, but not validate the legitimation of the value of the reported data. Thus, if the attacker is from legitimate but compromised nodes, it is very difficult for them to detect and distinguish the attacker from a normal reporter. Second, in mobile sensing systems that have high mobility, there is no permanent relationship between any two sensor nodes, so it is very difficult to verify each other by using the traditional security strategies like mutual authentication. In addition, the extreme large scale of the system and high mobility put a big challenge in key distribution if a security based approach is adopted. Finally, in such a totally distributed environment, all decisions should be made locally. Without the help of a central server, the deceptive data is detected only based on partial local information, which makes the problem more difficult. The difficulty is increased when the decision have to be made in a real-time way.

Besides security technologies, reputation based approaches, which usually require strong identity, cannot work in this case because of the possible large scale, the highly dynamics, and lack of the help of a center server in the system. Several other previous efforts have also been made in deceptive data detection and cleaning; however, most of them assume a specific distribution of the monitoring parameter, and they use this distribution as a model to predict and check the reported value of monitoring parameters. These methods can be useful techniques to detect deceptive data, but they rely a lot on the correctness of the distribution, so they cannot be generally extended to many applications. In summary, we find that deceptive data detection and filtering is the key to improve the quality of the collected data and the performance of dynamic sensor systems. With the insufficient of previous solution, it is essential to propose novel effective solutions to detect and filter deceptive data.

9.2 A Framework to Detect Deceptive Data

It is necessary to clean deceptive data to improve the system performance of wireless sensing systems. In this section, we design a framework, D^4, to detect and filter deceptive data, as shown in Figure 9.1. shows the components in the framework, A total of four mechanisms are proposed in the D^4 framework to detect two types of deceptive data as much as possible, in which, we propose four mechanisms to detect and filter deceptive data. Some mechanisms can be used in detecting and filtering both deceptive data, while some of them are specifically designed for only one type of deceptive data, either the redundant data or the false data.

From the figure, we can see that the process of deceptive data detection consists of two steps. The first step is done locally at each sensor that gets access to the data, when the data is sampled or received during transmission, and the second step is executed at the sink, when a lot of data has been delivered to the sink. At the first step, three detection mechanisms are defined, including *quality-assured aggregation and compression, self-learning model-based detection*, and *role-differentiated cooperative deceptive detection and filtering*, among which we think that the self-learning model-based detection can be used to detect and filter both types of deceptive data, although the parameters in the model may be adjusted to achieve a optimized result. Quality-assured aggregation and compression and role-differentiated cooperative detection are devised for detecting and filtering redundant data and false date independently. After all data arrive at the sink, the second step is conducted. A general spatial-temporal data quality check will be applied to check the

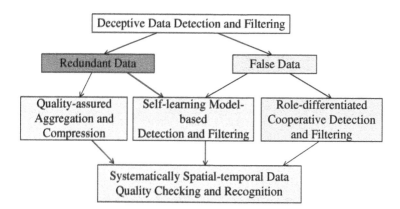

Figure 9.1: An overview of the framework to detect and filter deceptive data.

consistency of the all collected data. Next, we give an overview of these four detection and filtering mechanisms one by one.

9.2.1 Quality-Assured Aggregation & Compression

Quality-assured aggregation and compression is designed to detect and filter redundant data. It trades off the amount of delivered data with the quality of the collected data as well as fault tolerance. This mechanism works as follows. First, the quality requirements of the monitoring parameters are pre-set at each sensor. Second, when the data is collected at each sensor, the sensor starts to process data aggregation and compression when necessary. In this case, due to the constraints of storage and communication bandwidth, the maximum amount of data that each sensor can store and send is limited. When the collected data does not exceed this limitation, the sensor stores all new coming data. When the sensor runs out of resource, but it senses or receives more new data, it will start the process of data aggregation and data compression. We define a similarity factor to represent the level of deference between two pieces of data. During the aggregation, two temporally continuous sensor reading with maximum similarity factor will be aggregated by removing one piece of sensor reading. This process continues until the amount of remaining data is within the storage limitation. Quality-assured aggregation and compression are very useful to be applied to filter redundant data, because each time the filtered data lose least information compared with others. Thus when the maximum amount of sensed data and the size of the available storage is fixed, the quality of the collected data can be assured with a high probability using our quality-assured aggregation and compression.

9.2.2 The RD^4 Mechanism

Role-differentiated cooperative deceptive data detection and filtering is devised to confirm the correct event reports and thus to prevent false event reports. In this mechanism, each sensing node is assigned a role, with corresponding right to confirm an event. When an event is detected at one sensor, the sensor starts to collect the confidence score of this event. The event will be confirmed and sent to other sensors when the confidence score exceeds a preset threshold. Otherwise, the event report will be dropped when confidence score expires after some time. When the event report is receive at another sensor, it will follow the same process to confirm and propagate the event report. The detail of this protocol is described in Section 9.3.

9.2.3 Self-Learning Model Based Detection

Self-learning model based detection and filtering is a mechanism to detect and filter both redundant data and false data. In this mechanism, models should be carefully defined to evaluate the legitimation of the collected data. Basically these models will explore the locally spatial-temporal relationship among the collected data by taking advantages of physical features of the sensing data. Because the decision will be made locally at each sensor or on the path from the data source to the sink, the models should be designed under the condition of limited storage and computation, as well as partial information. In most applications, the detection and filtering algorithm is also required to be real-time, because redundant data has to be dropped immediately to reduce resource usage and false data that will result in wrong system functions has to be filtered as early as possible to avoid disasters. To achieve this goal, we define some consistency models, which consists of a set of rules to check the quality of the collected data. To be efficient, these rules will be simple, so the overhead of the consistency checking based on these rules can be controlled. Also, the rule set is a flexible set, which may increase or be modified based on the experience of the sensors. If the collected data can pass the checking of the rule set, we regard it as not deceptive data, otherwise, it will be detected as deceptive data and be filtered. We adopt a self-learning mechanism to manage the size of the rule set. At first, only basic rules will be defined to check the consistency of collected data. With the sensors getting more experience in the data collection, the rule set can be extended. For example, if a piece of false data is not detected by sensor A, but it is detected and confirmed by other sensors, sensor B, after sensor A is informed this missing detection, it can modify the rule set by adding new rules or changing the parameter in the current rule. Thus, next time, when this type of false data arrives at sensor A, sensor A can detect it with a high probability.

9.2.4   Spatial-Temporal Data Recognition

After the data is collected at the sink, a systematically spatial-temporal data quality checking and recognition is executed. Because the sink is usually much more powerful than other sensors, more complicated consistency checking operations can be processed. With a large amount of data available at the sink, the spatial-temporal relationship among sensor readings can be examined based on pre-defined consistency models. For example, based on the analysis of the previous collected sensing data, we can abstract some distribution of the monitoring parameters. Thus, if one piece of the collected data is far away from the distribution, it must be a piece of false data. Moreover, although we try to filter redundant data, there will still be spatial redundant data because sensors' monitoring field may overlap and an event can be detected and delivered by several sensors independently. We observe that those redundant data usually exhibits a certain type of similarity, i.e., they are close in the value or reporting a similar event. In general, this type of redundancy offers us a chance to verify the legitimation of sensing data by mutual verification. Finally, we argue that several monitoring parameters may have relationships among them. For instance, the detected vehicle speed is closely related with the density of the vehicles on the road. When a vehicle reports a high driving speed and another sensor reports a high density of traffic, most probably there will be false reports, i.e., either the vehicle is reporting a false speed or the density of the traffic is not so high. We should take advantage of the relationship among different parameters so that we can model this type of features, and use those models to detect as well as filter more false data which is not able to be detected and filtered locally during the data is sensed and transmitted due to the availability of only partial information, limited computation and storage resources.

9.2.5 Discussion

Deceptive data detection and filtering is a challenge in wireless sensing system design. The framework we proposed in this dissertation provides four general mechanisms to achieve the goal of deceptive data detection and filtering; however, as we have known that wireless sensing systems are mostly application specific systems. Thus, for specific wireless sensing system applications, we need to give details of each suitable mechanism and adapt our mechanism to satisfy the specific application requirements in detecting and filtering deceptive data so that the deceptive data can be detected and filtered with a high probability. We argue that it is very difficult to guarantee detecting and filtering all deceptive data, but we can make efforts in increasing the probability that these data is detected and filtered. In this framework, we totally introduce four mechanisms, but we will only give details of the RD^4 mechanism and leave the rest of them as our future work.

9.3 Role-differentiated Cooperative Deceptive Date Detection and Filtering

Among the four proposed mechanisms to detect and filter deceptive data, in this dissertation, we try to address one of them in detail, which is the RD^4 mechanism . In this section, we describe the detail of the mechanism as follows.

In the mechanism of role-differentiated cooperative deceptive data detection and filtering (RD^4), when a sensor is deployed, it picks up a role from the role set based on the specific features of the sensor, such as storage size, computation ability, communication ability, and trustable level, which specifies how much can this type of sensor be trusted. Then, each sensor plays a different role in the system and has the functions assigned to that role. The role set of a specific wireless sensing system application can be pre-defined as $R = \{R_1, R_2, R_3, ..., R_i\}$. For each event it sensed or received, the sensor can issue a confidence score to the event, which is denoted as $csr(E, T)_{ij}$, denoting the truth level of this event, where E specifies the event, and i and j are the identity of the role and the identity of the sensor independently, while T means the score will be valid for T time slots. We define the maximum confidence score that a sensor with ID j and role R_i can issue to a piece of data or an event report as $CSR(E, T)_{ij}$, which should satisfy $csr(E, T)_{ij} \leq CSR(E, T)_{ij}$.

In the RD^4 mechanism, whenever an event report is generated or received at a sensor, the sensor will check whether it is a false event or not. Thus, an event report will be checked the whole way from location it is detected to the sink of that event. Next we give details on how to set the confidential score and how to detect and filter a false event.

In RD^4, the confidential score defined above is calculated based on the accumulated signal strength during a certain time period $([0, T_0])$, depicted as $ASS(E, T_0)_{ij}$, of the corresponding event E at sensor j with role i. Here, the signal strength of an event E, denoted as $SS(E)_{ij}$, can be defined as the amount of changes of a monitoring physical parameter within a unit time period. For example, if we try to detect an event of sudden changes in temperature at a computing node in a high performance computing system. The signal strength will be the amount of temperature changing within each minute. Of course, for different types of application, the signal strength can be defined in different ways based on the special physical parameters. Thus, if the function to specify the changing rate of a monitoring physical parameter is $p(t)$, we can define $SS(E)_{ij}$ as

$$SS(E)_{i,j} = p(t)dt$$

Based on the defined $SS(E)_{ij}$, the accumulated signal strength can be defined as

$$ASS(E, T_0)_{ij} = \int_0^{T_0} SS(E)_{ij} = \int_0^{T_0} p(t)dt$$

Having the $ASS(E, T_0)_{ij}$, we design a function f that maps the accumulated signal strength to a confidential score, to be specific, $csr(E, T)_{ij} = f(ASS(E, T_0)_{ij})$. When an event is detected at a sensor, the sensor will set up a timer T, also used as the first lifetime period of the event, to the detected event. Then the sensor will try to confirm where it is a really event or just a faked one before the event expires. The decision is made based on the confidence score of the event, which can come from two sources. One is the observation by the sensor itself, for which we use accumulated signal strength detected by the senor, and the other is the reported signal strength about the same event from other sensors. Then the sensor j assigns a confidence score, $csr(E, T)_{ij}$, to the detected event, E, within a lifetime of T as follows.

$$csr(E, T)_{ij} = f(ASS(E, T_0)_{ij})$$

$$= \begin{cases} ASS(E, T_0)_{ij} & ASS(E, T_0)_{ij} \leq CSR(E, T)_{ij} \\ CSR(E, T)_{ij} & ASS > CSR(E, T)_{ij} \end{cases}$$

Based on the above formula, if the calculated confidence score exceeds the bound which is pre-set by the application to confirm an event. The sensor that detects the event will generate an event report and broadcast the event report to all other sensors in the sensing systems. The event report, denoted as $R(E)$, is a five tuple, including the information about the signal strength of the event, the first lifetime period of the event, event ID, event detector ID and the role of the detector. Thus, we have $R(E) = (AAS(E, T_0)_{ij}, T, E, j, i)$. Note that T is the lifetime of the event and T_0 is the latest timestamp when the event is seen. With the event ID and the detector ID, the event can be uniquely determined. Otherwise, the sensor will wait to collect more confidence score within the lifetime of the event, or drop the event when the event expires. When the event report is propagated, the report will be received by other sensors. If we assume the k_{th} sensor receive a set of event reports from its N neighbors, we can calculate the accumulated signal strength for event E as follows.

$$ASS(E, T_0)_{rk} = \sum_{n=0}^{N} W_{jn} * ASS(E, T_0)_{jn} + \int_{0}^{T_0} SS(E)_{rk}$$

where $ASS(E, T_0)_{lk}$ depicts the accumulated signal strength of the event E at sensor K with role r, which consists of two parts, the aggregation of received signal strength of event E from other sensors, denoted by $ASS(E, T_0)_{jn}$, and the accumulated signal strength based on its own observation at sensor k. W_{jn} is the weight of the event report from sensor n. It can be defined as the reverse of the distance between two sensors, the similarity of the two events detected by the two sensors, the trustable level of the sensor n, or the importance of the sensor n to sensor k. We argue that the definition of the weight should be adaptive to the nature of different sensing systems applications as well as requirements of the sensing systems applications.

Having the calculated accumulated signal strength, we can calculate the confidence score at sensor k. Based on the confidence score, we can make a judge on the truth of the event. That is to say, If the confidence score exceeds the pre-set bound to confirm the event, the event is confirmed and a new report about this event is generated and forwarded. Otherwise, it will wait another T more time period to check the accumulated signal strength and calculate the new confidence score again. Considering the timeliness of events, in our design, the signal strength will degrade with the time passing. Thus, at the end of each T time period, signal strength recalculated based on the following formula.

$$ASS(E, T_0 + T)_{rk} = \alpha ASS(E, T_0)_{rk}$$

$$+ \sum_{n=0}^{N} W_{jn} * ASS(E, T_0 + T)_{jn} + \int_{T_0}^{T_0+T} SS(E)_{rk}$$

In the above formula, $ASS(E, T_0 + T)_{rk}$ depicts the accumulated signal strength at sensor k against event E at the time $T_0 + T$. α is the degrading rate of the signal strength. Thus, in the formula, the first part is the degraded signal length observed before time T_0, and the second part is the received accumulated signal strength from other sensors between time T_0 and time $T_0 + T$; while the third part is the accumulated signal strength observed by sensor k in the time interval $[T_0, T_0+T]$. The event will be regarded as a false event after it is checked twice but does not confirmed yet. In other words, the event will be discard if it cannot be confirmed within $2T$. Based on the confidence score, a final judgement on the truth of the event is generated as follows.

$$Valid(E) = \begin{cases} True & csr(E, T)_{ij} > \theta || csr(E, 2T)_{ij} > \theta \\ False & otherwise \end{cases}$$

This formula means that the event is confirmed as true when confidence score exceeds a pre-set threshold; otherwise, the event is confirmed as false and the propagation of the event report will be terminated. In this way, we can image that the true event will be propagated very fast and it will be dropped after a while either because the farther sensor cannot detect the event or not so many sensor are sensing the reports. Thus, we can control the propagation of the event within a reasonable big area. If the event is a faked event, then it will have very little chance to be confirmed and propagated, so the faked events can be controlled within a very small area.

We argue that waiting $2T$ periods to confirm a reported event has advantages over just waiting one T period. First, if we only wait T period to confirm the even5, some true event may be dropped very earlier because not enough signal strength has been collected, which will be true in a highly distributed sensing environment with only insufficient local information. Of course, we can wait several time intervals of T to get enough strong signal strength; however, because the event report has timely features, the signal strength will degrade after some time the event is detected and reported. In this case, the number of waiting period, nT, should be controlled and the degrading of the signal strength should be taken into consideration as well. Finally, some wireless sensing applications have high real-time requirements, so we cannot make the waiting period too long.

In this section, we propose a general mechanism, role-differentiated cooperative detection and filtering, for sensing systems to detect and filter false data, especially false event reports. Because variant sensor network applications have their own features, our mechanism should be adjusted to fit the application requirements of specific applications. In this dissertation, we try to verify the effectiveness and efficacy of our mechanism in a dynamic sensor network environment, thus we use a vehicular network as an example to test our mechanism. In the next section, we give details about how to use our mechanism to detect and filter deceptive data in the context of vehicular networks.

9.4 RD^4 in Vehicular Networks

We propose a general mechanism using role-differentiated cooperative approach to detect deceptive data. We argue that our mechanism can be applied in a lot of applications, however, we will only evaluate the efficiency and efficacy of our mechanism in one application, which is a dynamic sensor network. As we have pointed out in previous section, our mechanism should be adjusted to satisfy application requirements of different applications. In this section, we give an example of how to adjust our role-differentiated cooperative deceptive data detection mechanism in a vehicular network application.

The RD^4 mechanism is a general mechanism to detect deceptive data. In this section, we adapt the RD^4 mechanism to detect false accident report in the context of vehicular networks.

9.4.1 Role Definition in Vehicular Networks

In this subsection, we define a detailed role-differentiated false accident event detection protocol in vehicular networks. The first step to define the RD^4 mechanism is to define a set of different roles in vehicular network applications. Considering different types of function components such as vehicles and road side units (RSUs) in vehicular networks, we classify those function components to four roles, including RSUs, public vehicles such as police cars, school buses and so on, regular vehicles like personal owned cars, and vehicle itself. Thus the role set in the vehicular networks is defined as $R = \{R_{rsu}, R_{pub}, R_{reg}, R_{self}\}$. For each role R_i, it can assign a maximum confidence score, CSR_{ij}, to an accident report it detests or confirms based on the characters of the role. In our design, the definition of maximum confidence score (CSR) is closely related the trustable level of each role in the vehicular network. For example, we think that RSUs are more trustable than all vehicles that are on the road because of two reasons. First, they are mostly controlled by public organizations such as Department of Transportation (DoT) or some certificated companies, thus it is more difficult for attackers to compromise these devices. Moreover, because the RSUs have more power in terms of computation, storage and communication, fixed-location deployment and relatively high availability, they usually have more information than other vehicles. As a result, RSUs usually reports more accurate data than other on-road vehicles so that they have higher trustable level, thus they can issue bigger value of confidential code and send out stronger signal strength. Comparing the public vehicles with regular vehicles, we argue that public vehicles have more protection than the regular ones, e.g., police cars are equipped with high security devices and always managed by police department, so public vehicles are more creditable than regular personal owned vehicles. Except those three regular roles, we have another special role, R_{self}, in our system, for which we believe that it should have the highest trustable level. Thus, we define the CSR_{ij}s following the order of $CSR_{self,j} > CSR_{rsu,j} > CSR_{pub,j} > CSR_{reg,j}$.

9.4.2 False Accident Report Detection

In this dissertation, we assume that the detection of a true accident is handled by the accident sources, which can either be the vehicles that are involved in the accident or the police cars that are taking care of the accident. This assumption is based on the following observations. When an accident happens, thought the vehicle will be damaged, even totally, not all parts in the vehicle will crash. For example, in most cases when an accident happens, at least two tires of the vehicle will still be in a good condition. Thus, the sensors equipped on the those two tires can initialize an accident report and propagate it to other passing vehicles. On the other hand, the accident report can also be initially generated by police cars. In most cases, when an accident appears, a police car will come to the site where the accident appears very soon. Except those two types of vehicles, other vehicles are not supposed to report an accident. In other words, any accident reports generated by vehicles except above two type are false accident reports. Malicious vehicles may insert false accident report by performing as the vehicles involved in the accident. Thus, our goal is to remove false accident reports from malicious vehicles. In this dissertation, we assume that each vehicle is equipped a tamper-proof component, so even if a vehicle is compromised, it still cannot generating multiple identities.

When an accident report is received by a vehicle on the road. The vehicle will make a judgement about the truth of the accident report, based on the signal strength from its own observation and the signal strength reporting the same event from other vehicles. In our design, based on the reality that traffic will be blocked so that the vehicles will slow down when an accident happens. We use the vehicle velocity deceleration as the signal strength defined in the model. Thus $p(t) = a(t) = dv/dt$, where $a(t)$ is the acceleration rate, and v is the velocity of the vehicle. Then, the signal strength observed by vehicle j, $ASS(E, T_0)_{ij}$, can be calculated as $ASS(E, T_0)_{ij} = \int_0^T a(t) = v_T - v_0 = \Delta v$. To emphasize the importance of roles, we assign important role a special ability to send out accident reports with stronger signal strength, thus after we calculate the accumulated signal strength observed by the vehicle itself, we will adjust the accumulated signal strength by considering the role of the vehicle. The new accumulated signal strength we get is

$$ASS(E, T_0)_{ij} = \frac{CSR(E, T)_{ij}}{CSR(E, T)_{reg,j}} \Delta v$$

Except the signals strength gained from the observation of the vehicle j, vehicle j will receive signal strength for the event E from other vehicles. If we integrate both types of signal strength of the event E, we get an integrated value of the accumulated signal strength of event E at vehicle j with timestampe T_0 as follows.

$$ASS(E, T_0)_{ij} = \sum_{n=0}^{N} W_{jn} * ASS(E, T_0)_{jn}$$

$$+ \frac{CSR(E, T)_{ij}}{CSR(E, T)_{reg,j}} \Delta v$$

Similarly, we can calculate the integrated value of the accumulated signal strength after a period of T time slots, which is specified as following.

$$ASS(E, T_0 + T)_{ij} = \alpha ASS(E, T_0)_{ij}$$

$$+ \sum_{n=0}^{N} W_{jn} * ASS(E, T_0 + T)_{jn} + \frac{CSR(E, T)_{ij}}{CSR(E, T)_{reg,j}} (v_{T_0+T} - v_{T_0})$$

Having calculated the integrated value of the accumulated signal strength, we can map it to get a confident score for the event E at the vehicle j using the same function we define in Section 9.3.

Considering the fact that an accident on the road will only affect the vehicles close the the area where the accident is detected, we give more weight to an accident report that is from a location closer to the accident. Also, we argue that reporting of accident is more trustable for the vehicles located closer to the sources, since the vehicles located closer to the source is more likely to observe the accident than farther ones. Thus we define the value of W_{jn} as the reverse of the distance between the location of reporting vehicles and the location of the accident source, i.e., $W_{jn} = \frac{1}{dst(n,a)}$, where $dst(n,a)$ specifies the distance between vehicle n and the accident.

Based on the value of the confident score, we can make a judgement on the truth of event E at vehicle j. If the accident is confirmed as a true accident, it will be propagated by sending a message from vehicle j, including the event report, $ASS(E,T)_{ij}, T, E, i, j$, where $ASS(E,T)_{ij}$ is the signal strength of the event E, while i and j is the role and the identity of the vehicle independently. Otherwise, the accident report will be dropped. For different vehicle roles in the networks, we may also assign different values of threshold. For example, the threshold of the RSUs will be higher than that of regular vehicles, because RSUs are more powerful than regular vehicles and they have more effect on the decision making than regular vehicles to other vehicles in the network.

9.5 Performance Evaluation

We are in a position to evaluate the effectiveness and efficiency of our proposed protocol. In this section, we will first describe a vehicular network simulator. Then, we will evaluate our protocol in terms of several performance metrics, namely recall, accident report propagation range, accident report confirmation time, respectively in the rest of the this section.

9.5.1 Simulation Setup

We design a simulator for vehicular networks, by extending the traffic simulator designed in [Thiemann et al., 2008, Traffic Simulator,] that simulate the movements, such as acceleration, decrepitation, exchange lane, of the vehicles. We simulate the RD^4 mechanism for vehicular network in a road segment of a two-lane one way highway scenario with a ramp where vehicles enter the highway. We extend the simulator is the following three ways.

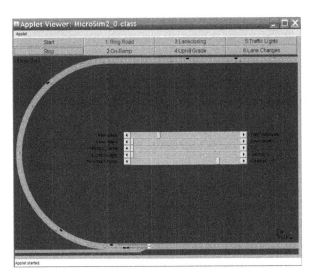

Figure 9.2: A snapshot of vehicular network simulator.

First, we implement a communication subsystem for the vehicular networks so that all components, including all vehicles and RSUs, in the network can communicate. In addition, we simulate the scenario of accident as well as malicious vehicles including how they use the mentioned communication system to broadcast fake or real alarm to their neighbors. Last but not the least, we fulfill the RD^4 mechanism to record and classify the report based on the description in Section 9.4. A snapshot of the simulator is shown in Figure 9.5.1, where the red dots depict the regular vehicles and the black dots denote the public vehicles. The accident is specified by the white block.

In our experiments, the communication between the vehicles follows the specification of DSRC using the maximum communication range 200 meters. For each vehicle the speed limitation is $75mph$. The road segment consists of a U-shape road with length of 6575 meters and a ramp. When an accident appears, the road will be closed for several minutes.

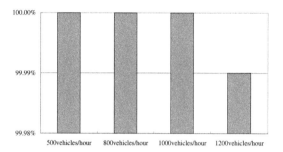

Figure 9.3: Recall of false accident reports.

The malicious vehicle will periodically broadcasts faked accident event if it doesn't detect a true event. Otherwise, it will keep silent.

9.5.2 Effectiveness of RD^4

In this subsection, we show how effective the RD^4 mechanism can detect the false accident report inserted by malicious vehicles and confirm the true accident report generated by the vehicles involved in the accident. This property is evaluated by the recall, which is defined as the fraction in the amount of a certain event reported that our mechanism classifies the report as this event. To be specific, in this application, the recall of false accident report is defined as the fraction of the false accident reports that is detected by the RD^4 mechanism, which is shown in Figure 9.5.2, while the recall of true accident report is defined as the fraction of true accident reports confirmed by the RD^4 mechanism, which is show in Figure 9.5.2. In this experiment, we set the value of the maximum confidential score, $CSR_{self}, CSR_{rsu}, CSR_{pub}$ and CSR_{reg} to be $5, 3, 2, 1$ respectively.

In the above two figures, the x-axis is the average vehicle density on the road and the y-axis shows the recall. we can easily observe that the RD^4 mechanism detects 100% false accident reports in most cases and more than 95% nearby vehicles confirm a true accident

154

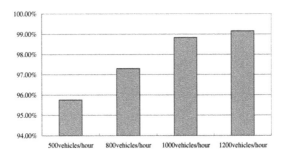

Figure 9.4: Recall of true accident reports.

report in all cases. These results validate that the proposed RD^4 mechanism is very effective in false accident report detection. We also find the the recall changes with the changing of the vehicle density on the road. The recall of false accident report drops a little when vehicle density grows, while the recall of true accident report increases with the increase of vehicle density. This is because more traffic likely brings a lower average speed, which is helpful to confirm a true accident in that vehicles are easier to get a low speed when an accident happens, but a low speed is also helpful to confirm any accident report, so it will affect the accuracy in detecting a false accident report contributes. Fortunately, based on our experiment, we find that the effect is big only when the speed of vehicles is very low, which depicts a heavy traffic jam and can be regards as a true accident. From this point of view, when traffic is low, a false accident report is helping us to broadcast the true traffic information.

9.5.3 Efficiency of RD^4

After we validate the effective of RD^4, we show the efficiency of RD^4. Given that almost all false accident reports will be detected and filtered, we only show how efficient RD^4 can

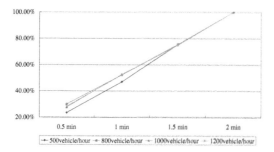

Figure 9.5: Confirmation time of true accident reports.

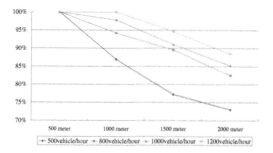

Figure 9.6: Propagation range of true accident reports.

confirm true accident reports. Basically, we evaluate how fast a true accident report can be confirmed and how far it can be propagated to be confirmed.

Figure 9.5.3 shows the percentage of nearby vehicles (within X meters) that a true accident report is confirmed. The x-axis depicts the time and the y-axis denotes the percentage of confirmation. Four lines in different colors show different scenario of different average vehicle density. We assume the total number of vehicles confirm the alarm in 2 minutes as entity. In the figure, as the density increases from 500 to 800 vehicle per hour, more vehicles detect the accident in the first half minutes due, which follows the same observation that high average vehicle density helps to confirm accident reports. With the time passing, the confirmation percentage increases almost linearly, approximately 25% every half minute, which is because RD^4 needs to collect sufficiency signal strength to confirm the accident report. The confirmation rate increases quickly with the accumulation of the signal strength, which may come from the fact of the slowing down vehicle velocity, the increasing average density, and the more propagating accident reports when there is a true accident.

The propagation of a true accident confirmation is depicted in Figure 9.5.3, where the x-axis records the distance from the vehicle to the accident location and the y-axis shows the percentage of vehicles of the corresponding distance that confirm the accident report. Similar to above experiment, four scenarios with different average vehicle densities are reported in the figure. We can see that with the increasing of density, for the same range, more percentage of vehicles detect the accident because high density will usually result in low velocity and more confirmation messages about the accident report. Those several factors interweave and produce this result. We can also see that more percentage of vehicles located close to the accident confirm the true accident report than the vehicles located faraway from the accident location. For example, when the density is more than 1000 vehicles per hour, more than 97% vehicles within the range of 1000 meters confirm the accident report, and the confirmation rate reduces to be between 85% and 90% when the vehicles are at the range of 2000 meters. This shows the degrading of the signal strength of the accident with the increasing of distance between the vehicles and the accident. Based on our simulation, even though the distance goes to more than 2000 meters, a lot of vehicles can still confirm the accident ahead on the road and they can take an early plan to switch the road. Furthermore, if the average vehicle density is higher or the communication range is larger, the accident confirmation can propagate to vehicles located further.

157

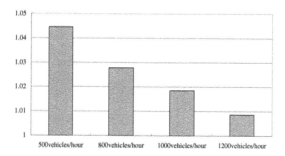

Figure 9.7: Message complexity of RD^4.

9.5.4 Message Complexity

Having seen the effectiveness and efficiency of the RD^4 mechanism, we evaluate the overhead of this mechanism. Figure 9.5.4 shows the number of messages delivered to confirm a true accident report in terms of different average vehicle densities. In the figure, we observe that the average number of messages send to "persuade" a vehicle is very small, to be specific, on average one message is enough. Actually, the cost only accumulates at the beginning of an accident, with more vehicles broadcast alarm reports, this cost barely increases. Also, as we expected, the message complexity decrease as the density grows. Because more vehicles broadcast alarm reports once they detect the accident, it costs fewer messages on average to convince a vehicle.

9.5.5 Effect of Maximum Confidential Scores

In the RD^4 mechanism, different roles are assigned different capability in issuing a confidential score. In this subsection, we try to tune the confidential score each role can issue and compare the results which helps us to understand this mechanism. All experiments are tested under traffic density of 1000 vehicles per hour.

CSRself	5	7	4	5	5	5	5	5	5
CSRrsu	3	3	3	4	2	3	3	3	3
CSRpub	2	2	2	2	2	3	0.5	2	1
CSRreg	1	1	1	1	1	1	1	2	0.5
Recall(false)	1	0.996	1	1	1	1	1	0.923	1
Recall(true)	0.988	0.990	0.988	0.989	0.988	0.989	0.998	0.994	0.968

Figure 9.8: Effect of maximum confidential scores.

Figure 9.5.5 shows the effect of tuning $CSRs$ to the recalls. Each data column in the figure reports a set of the maximum $CSRs$ of four roles and the resulted recalls using this set of $CSRs$. The first data column is the set of the maximum $CSRs$ used in experiments in above several subsections. From the figure, we can see that tuning the CSR_{rsu} and CSR_{pub} will not change the recall too much. This is because in our simulation the number of RSUs and public vehicles are very small, which is close to the reality, but we still see a slight improvement in the recall of true accident report when CSR_{rsu} and CSR_{pub} are assigned higher value, which reflects the fact that those roles are worthy to be trusted. Tuning the value of CSR_{self} affects both recalls. If we increase the value of CSR_{self}, the vehicles become more sensitive, thus they will confirm more true accident reports as well as more false accident reports. Thus, to detect and filter the false accident report, we should not make the vehicles too sensitive in confirming events.

Compared with above three $CSRs$, tuning CSR_{reg} affects the recalls most. When we increase the value of CSR_{reg}, the recall of false accident reports degrades a lot but the recall of true accident reports increases, because of the large amount of regular vehicles in the system and higher probability of regular vehicles to be malicious vehicles than other types of components in the system. Similarly, when we decrease the value of CSR_{reg}, less true accident reports are confirmed, but more false accident reports are confirmed. Thus, we need to find a good value for CSR_{reg} to balance the recall of false accident reports and the recall of true accident reports.

In summery, we find that the RD^4 mechanism works effectively and efficiently in detecting and filtering false accident reports as well as confirming true accident reports at a low cost. A tradeoff should be made to assign picking maximum confidential score for each role.

CHAPTER 10

CONCLUSIONS AND FUTURE WORK

10.1 Conclusion

In this dissertation, we we envision that the success of the sensing system applications are largely depending on whether we can collect high quality data using the deployed sensing systems. Thus, we propose to integrate data quality management in any energy efficient sensing system design. As a result, a general framework named *Orchis* has been proposed to manage data quality in sensing systems. In the framework, we first analyze the characteristics of the sensing data. Then, we propose a set of data consistency models that are the core of our framework and are used to evaluate the quality of the collected data. Moreover, we devise a set of protocols to achieve the goals of both saving energy and collecting high quality data. For example, an energy efficient adaptive lady protocol is designed to collect high quality data in an energy efficient way. To improve the data quality of the collected data, we also propose a systematic mechanism to detect and filter the deceptive data. Finally, we argue that the protocol in sensing systems should be energy efficient, so we formally define a set of models, lifetime models for sensor networks, to evaluate the energy efficient of the proposed protocol. According to the results from both the simulation based on more than 100 sensors using TOSSIM and the prototype implementation based on 13 Mica2 Motes, we validate the effectiveness of our proposed protocol.

10.2 Future Work

After having introduced the work we have done in this dissertation, we are in a position to depict what we plan to do in future research.

10.2.1 Complete and Extend the Orchis Framework

We plan to completing some undeveloped protocols in the design of the framework, such as several other mechanisms to detect and filter deceptive data and traffic-differentiated cross-layer protocol. Then we will extend the Orchis framework to fit Cyber-Physical Systems, for example, consistency models can take advantage of the features of the physical properties. In other words, to including some physical laws into the consistency models will be very helpful in consistency checking as well as deceptive data detection. Moreover, the system protocols will also be adjusted to fit the characteristics of cyber-physical systems accordingly. Mechanisms that are used to replace the detected deceptive data and missing data will be explored. In the case when we cannot avoid dropping some sensing data due to the constrained computation, communication and storage, we will drop less important data based on the physical characteristics of the data. Furthermore, cross-layer traffic-differentiated data collection protocols will be designed to transmit data in physical priority ways. Finally, we will implement and evaluate more protocols we mention in the deceptive data detection and filtering framework.

10.2.2 Privacy and Security in Sensing Systems

In addition, Security and privacy are always of importance for modern systems. In cyber-physical systems or extended sensing systems, the deep distributed physical environment both increases the risk and offers new opportunities. For example, sensitive biomedical data collected from body area networks may release critical personal information. A security breach may result in fatal physical damage, such as vehicle crash. Thus, it is a big challenge to preserve privacy and security in such an open environment; whereas, the distinct physical features can be also utilized to authenticate legislate users and detect malicious attackers. In this direction, I plan to extend the concept of adaptive privacy and propose novel protocols to preserve privacy and improve security.

10.2.3 Sensing System Applications in Healthcare Systems

Finally, Cyber-physical systems and the extended sysnsing systems can be used in a great deal of applications. One of the most promising applications is in developing modern health-self-care systems. We proposed a Smart Phone Assisted Chronic Illness Self-Management System in [Sha et al., 2008c], which uses participatory sensing to investigate the causes of chronic illness and discover potential approaches to prevent chronic illness. In this system, all above proposed techniques will be integrated in a prototype system and they will be evaluated in real deployment.

REFERENCES

[Adam et al., 2004] Adam, N., Janeja, V. P., and Atluri, V. (2004). Neighborhood based detection of anomalies in high dimensional spatio-temporal sensor datasets. In *Proceedings of the 2004 ACM symposium on Applied computing.*

[Akkaya and Younis, 2003] Akkaya, K. and Younis, M. (2003). An energy-aware qos routing protocol for wireless sensor networks. In *Proceedings of the IEEE Workshop on Mobile and Wireless Networks (MWN'03).*

[Akyildiz et al., 2002] Akyildiz, I. F., Su, W., Sankarasubramaniam, Y., and Cayirci, E. (2002). A survey on sensor networks. *IEEE Communications Magazine*, 40(8):102–114.

[Arisha et al., 2002] Arisha, K., Youssef, M., and Younis, M. (2002). Energy-aware tdma-based mac for sensor networks. In *IEEE IMPACCT 2002.*

[Arora et al., 2005] Arora, A. et al. (2005). Exscal: Elements of an extreme scale wireless sensor network. In *Proceedings of the 11th IEEE International Conference on Embedded and Real-Time Computing Systems and Applications (RTCSA 2005).*

[Atkins, 2006] Atkins, E. M. (2006). Cyber-physical aerospace: Challenges and future directions in transportation and exploration systems. In *NSF Workshop on Cyber-Physical Systems.*

[Bao and J.J.Garcia-Luna-Aceves, 2001] Bao, L. and J.J.Garcia-Luna-Aceves (2001). Channel access scheduling in ad hoc networks with unidirectional links. In *International Workshop on Discrete Algorithms and Methods for Mobile Computing and Communications (DIALM'01).*

[Batalin et al., 2004] Batalin, M. et al. (2004). Call and responses: Experiments in sampling the environment. In *Proc. of ACM SenSys 2004*.

[Bhardwaj and Chandrakasan, 2002] Bhardwaj, M. and Chandrakasan, A. (2002). Bounding the lifetime of sensor networks via optimal role assignments. In *Proceedings of IEEE INFOCOM 2002*.

[Bhardwaj et al., 2001] Bhardwaj, M., Garnett, T., and Chandrakasan, A. P. (2001). Upper bounds on the lifetime of sensor networks. In *Proceedings of ICC 2001*.

[Bharghavan et al., 1994] Bharghavan, V., Demers, A., Shenker, S., and Zhang, L. (1994). Macaw: A media access protocol for wireless lans. In *Proc. of ACM SIGCOMM'94*.

[Bishop, 2000] Bishop, R. (2000). A survey of intelligent vehicle applications worldwide. In *Proc. of IEEE intelligent Vehicles Symposium 2000*.

[Blough and Santi, 2002] Blough, D. M. and Santi, P. (2002). Investigating upper bounds on network lifetime extension for cell-based energy conservation techniques in stationary ad hoc networks. In *Proceedings of the 8th Annual ACM/IEEE International Conference on Mobile Computing and Networking(MobiCom'02)*.

[Bokareva et al., 2006] Bokareva, T. et al. (2006). Wireless sensor networks for battlefield surveillance. In *Proceedings of Land Warfare Conference 2006*.

[Braginsky and Estrin, 2002] Braginsky, D. and Estrin, D. (2002). Rumor routing algorithm for sensor networks. In *Proceedings of the First Workshop on Sensor Networks and Applications(WSNA'02)*.

[Campbell et al., 2006] Campbell, A. et al. (2006). People-centric urban sensing. In *Proceedings of the 2nd Annual International Workshop on Wireless Internet*.

[Cerpa and Estrin, 2002] Cerpa, A. and Estrin, D. (2002). ASCENT: Adaptive self-configuring sensor network topologies. In *Proc. of IEEE Conference on Computer Communications (INFOCOM'02)*.

[Chang and Tassiulas, 2000] Chang, J. H. and Tassiulas, L. (2000). Maximum lifetime routing in wireless sensor networks. In *Proceedings of Advanced Telecommunications and Information Distribution Research Program(ARIRP'00)*.

[Chen et al., 2001] Chen, B., Jamieson, K., Balakrishnan, H., and Morris, R. (2001). SPAN: An energy-efficient coordination algorithm for topology maintenance in ad-hoc wireless networks. In *Proceedings of the 7th Annual ACM/IEEE International Conference on Mobile Computing and Networking(MobiCom'01)*.

[Chipara et al., 2005] Chipara, O., Lu, C., and Roman, G. (2005). Efficient power management based on application timing semantics for wireless sensor networks. In *Proceedings of the 25th International Conference on Distributed Computing Systems (ICDCS'05)*.

[COMPASS,] COMPASS. Compass: Collaborative multiscale processing and architecture for sensornetworks.

[crossbow,] crossbow. Crossbow technology inc.

[Dai and Han, 2003] Dai, H. and Han, R. (2003). A node-centric load balancing algorithm for wireless sensor networks. In *Proceedings of IEEE GLOBECOM'03*.

[Dam and Langendoen, 2003] Dam, T. and Langendoen, K. (2003). An adaptive energy-efficient mac protocol for wireless sensor networks. In *Proc. of ACM SenSys 2003*.

[Dasgupta et al., 2003] Dasgupta, K. et al. (2003). An efficient clustering-based heuristic for data gathering and aggregation in sensor networks. In *Proceedings of the IEEE Wireless Communications and Networking Conference (WCNC'03)*.

[Datta, 2003] Datta, A. (2003). Fault-tolerant and energy-efficient permutation routing protocol for wireless networks. In *International Parallel and Distributed Processing Symposium (IPDPS'03)*.

[Deb et al., 2002] Deb, B., Bhatnagar, S., and Nath, B. (2002). A topology discovery algorithm for sensor networks with applications to network management. In *In IEEE CAS workshop(short paper)*.

[Deb et al., 2003] Deb, B., Bhatnagar, S., and Nath, B. (2003). Information assurance in sensor networks. In *Proceedings of 2nd ACM International Workshop on Wireless Sensor Networks and Applications*.

[Ding et al., 2003] Ding, M., Cheng, X., and Xue, G. (2003). Aggregation tree construction in sensor networks. In *Proceedings of the 58th IEEE Vehicular Technology Conference'03*.

[Douglas et al., 2006] Douglas, C. et al. (2006). Cyber-physical systems and wildland fire or contaminant identification and tracking dynamic data-driven application systems. In *NSF Workshop on Cyber-Physical Systems*.

[DSRC,] DSRC. Dedicated short range communications (dsrc) home.

[Du et al., 2005] Du, J., Shi, W., and Sha, K. (2005). Asymmetry-aware link quality services in wireless sensor networks. In *Proceedings of the 2005 IFIP International Conference on Embedded and Ubiquitous Computing*.

[Duarte-Melo and Liu, 2003] Duarte-Melo, E. and Liu, M. (2003). Analysis of energy consumption and lifetime of heterogeneous wireless sensor networks. In *in Proc. IEEE Globecom*.

[Elnahrawy and Nath, 2003] Elnahrawy, E. and Nath, B. (2003). Cleaning and querying noisy sensors. In *Proceedings of the Second ACM Workshop on Wireless Sensor Networks and Applications (WSNA)*.

[Elson et al., 2002] Elson, J., Girod, L., and Estrin, D. (2002). Fine-grained network time synchronization using reference broadcasts. In *Proc. of the Fifth Symposium on Operating Systems Design and Implementation(OSDI)*.

[Estrin et al., 2002] Estrin, D., Culler, D., Pister, K., and Sukhatme, G. (2002). Connecting the physical world with pervasive networks. *IEEE Pervasive Computing*, 1(1):59–69.

[Estrin et al., 1999] Estrin, D. et al. (1999). Next century challenges: Scalable coordination in sensor networks. In *Proceedings of the 5th Annual ACM/IEEE International Conference on Mobile Computing and Networking (MobiCom'99)*.

[Estrin et al., 2003] Estrin, D., Michener, W., and Bonito, G. (2003). Environmental cyberinfrastructure needs for distributed sensor networks. A Report From a National Science Foundation Sponsored Worksop.

[Faruque and Helmy, 2004] Faruque, J. and Helmy, A. (2004). RUGGED: Routing on fingerprint gradients in sensor networks. In *Proceedings of IEEE Int'l Conf. on Pervasive Services*.

[Ganesan et al., 2002] Ganesan, D. et al. (2002). Highly resilient, energy efficient multipath routing in wireless sensor networks. *Mobile Computing and Communications Review(MC2R)*, 1(2).

[Garces and Garcia-Luna-Aceves, 1997a] Garces, R. and Garcia-Luna-Aceves, J. J. (1997a). Collision avoidance and resolution multiple access: First-success protocols. In *Proceedings of IEEE International Conference on Communications (ICC'97)*.

[Garces and Garcia-Luna-Aceves, 1997b] Garces, R. and Garcia-Luna-Aceves, J. J. (1997b). Collision avoidance and resolution multiple access with transmission groups. In *Proc. of IEEE Conference on Computer Communications (INFOCOM'97)*.

[Garrett et al., 2006] Garrett, J., Moura, J., and Sanfilippo, M. (2006). Sensor-data driven proactive management of infrastructure systems. In *NSF Workshop on Cyber-Physical Systems*.

[Gedik and Liu, 2005] Gedik, B. and Liu, L. (2005). Location privacy in mobile systems: A personalized anonymization model. In *Proc. of the 25th International Conference on Distributed Computing Systems*.

[Gnawali et al., 2006] Gnawali, O. et al. (2006). The tenet architecture for tiered sensor networks. In *Proc. of ACM SenSys 2006*.

[Goel and Estrin, 2003] Goel, A. and Estrin, D. (2003). Simultaneous optimization for concave costs: Single sink aggregation or single source buy-at-bulk. In *Proceedings of the 14th Annual ACM-SIAM Symposium on Discrete Algorithms'03*.

[Goodman et al., 1989] Goodman, D., Valenzuela, R., Gayliard, K., and Ramamurthy, B. (1989). Packet reservation multiple access for local wireless communications. *IEEE Transactions on Communications*, 37(8):885–890.

[Gupta and Younis, 2003] Gupta, G. and Younis, M. (2003). Fault-tolerant clustering of wireless sensor networks. In *in the Proceedings of the IEEE Wireless Communication and Networks Conference (WCNC 2003)*.

[Gupta et al., 2008] Gupta, H. et al. (2008). Efficient gathering of correlated data in sensor networks. *ACM Transactions on Sensor Networks*, 4(1).

[Gupta et al., 2005] Gupta, R., Puri, A., and Ramamritham, K. (2005). Executing inco-
herency bounded continuous queries at web data aggregators. In *Proc. of 14th International World Wide Web Conference.*

[Haeberlen et al., 2004] Haeberlen, A. et al. (2004). Practical robust localization over large-
scale 802.11 wireless networks. In *Proceedings of the 10th Annual ACM/IEEE International Conference on Mobile Computing and Networking(MobiCom'04).*

[Hamdaoui and Ramanathan, 2006] Hamdaoui, B. and Ramanathan, P. (2006). *Energy-
Efficient and MAC-Aware Routing for Data Aggregation in Sensor Networks.* IEEE Press.

[He et al., 2002] He, T., Blum, B., Stankovic, J. A., and Abdelzaher, T. F. (2002). Aida:
Adaptive application independent data aggregation in wireless sensor networks. *ACM
Transactions on Embeded Computing Systems*, 40(8):102–114.

[He et al., 2003] He, T., Stankovic, J. A., Lu, C., and Abdelzaher, T. (2003). Speed: A
stateless protocol for real-time communication in sensor networks. In *Proceedings of
IEEE ICDCS'03.*

[Heinzelman et al., 2002] Heinzelman, W., Chandrakasan, A., and Balakrishnan, H. (2002).
An application-specific protocol architecture for wireless microsensor networks. *IEEE
Transactions on Wireless Communications*, 1(4):660–670.

[Heinzelman et al., 1999] Heinzelman, W., Kulik, J., and Balakrishnan, H. (1999). Adaptive protocols for information dissemination in wireless sensor networks. In *Proceedings
of the 5th Annual ACM/IEEE International Conference on Mobile Computing and Net-
working(MobiCom'99)*, pages 174–185.

[Heinzelman et al., 2000] Heinzelman, W. R., Chandrakasan, A., and Balakrishnan, H.
(2000). Energy-efficient communication protocol for wireless micorsensor networks. In
Proceedings of HICSS'00.

[Hill et al., 2000] Hill, J., Szewczyk, R., Woo, A., Hollar, S., Culler, D., and Pister, K. (2000). System architecture directions for networked sensors. In *Proceedings the 9th ASPLOS'00*, pages 93–104.

[Hodge and Austin, 2004] Hodge, V. and Austin, J. (2004). A survey of outlier detection methodologies. *Artificial Intelligence Review*, 22:85–126.

[IEEE802.11, 1999] IEEE802.11 (1999). Part 11: Wireless lan medium access control (mac) and physical layer (phy) specification.

[IEEE802.15.4/D18, 2003] IEEE802.15.4/D18 (2003). Draft standard: Low rate wireless personal area networks.

[Intanagonwiwat et al., 2002] Intanagonwiwat, C., Estrin, D., Govindan, R., and Heidemann, J. (2002). Impact of network density on data aggregation in wireless sensor networks. In *Proceedings of IEEE ICDCS'02*.

[Intanagonwiwat et al., 2000] Intanagonwiwat, C., Govindan, R., and Estrin, D. (2000). Directed diffusion: A scalable and robust communication paradigm for sensor networks. In *Proceedings of the 6th Annual ACM/IEEE International Conference on Mobile Computing and Networking(MobiCom'00)*.

[ITSA and DoT, 2002] ITSA and DoT (2002). National intelligent transportation systems program plan: A ten-year vision. A Report from Intelligent Transportation Society of America and Departemnt of Transportation.

[Jain and Chang, 2004] Jain, A. and Chang, E. (2004). Adaptive sampling for sensor networks. In *Proc. of the 1st international workshop on Data management for sensor networks: in conjunction with VLDB 2004*.

[Janakiram et al., 2006] Janakiram, D., Reddy, A., and Kumar, A. (2006). Outlier detection in wireless sensor networks using bayesian belief networks. In *Proceeedings of the 1st International Conference on Communication System Software and Middleware*.

[Jeffery et al., 2006] Jeffery, S. et al. (2006). Declarative support for sensor data cleaning. In *Proceeedings of The 4th International Conference on Pervasive Computing*.

[Kalpakis et al., 2002] Kalpakis, K., Dasgupta, K., and Namjoshi, P. (2002). Maximum lifetime data gathering and aggregation in wireless sensor networks. In *Proceedings of IEEE Networks'02 Conference(NETWORKS02)*.

[Kang and Poovendran, 2003] Kang, I. and Poovendran, R. (2003). Maximizing static network lifetime of wireless broadcast adhoc networks. In *IEEE 2003 International Conference on Communications*.

[Karp and Kung, 2000] Karp, B. and Kung, H. T. (2000). Gpsr: greedy perimeter stateless routing for wireless networks. In *Proceedings of the 6th Annual ACM/IEEE International Conference on Mobile Computing and Networking(MobiCom'00)*.

[Khanna et al., 2004] Khanna, G., Bagchi, S., and Wu, Y. (2004). Fault tolerant energy aware data dissemination protocol in sensor networks. In *2004 International Conference on Dependable Systems and Networks (DSN'04)*.

[Krause et al., 2008] Krause, A. et al. (2008). Toward community sensing. In *Proc. of the fourth International Conference on Information Processing in Sensor Networks (IPSN'08)*.

[Larkey et al., 2006] Larkey, L., Bettencourt, L., and Hagberg, A. (2006). In-situ data quality assurance for environmental applications of wireless sensor networks. Technical Report Report LA-UR-06-1117, Los Alamos National Laboratory.

[Lazaridis et al., 2004] Lazaridis, I. et al. (2004). Quasar: quality aware sensing architecture. *ACM SIGMOD Record, Special section on sensor network technology and sensor data management*, 2:26–31.

[Lazaridis and Mehrotra, 2003] Lazaridis, I. and Mehrotra, S. (2003). Capturing sensor-generated time series with quality guarantees. In *Proceedings of 19th International Conference on Data Engineering*.

[Levis et al., 2003] Levis, P., Lee, N., Welsh, M., and Culler, D. (2003). Tossim: Accurate and scalable simulation of entire tinyos applications. In *Proc. of ACM SenSys 2003*.

[Li et al., 2007] Li, H. et al. (2007). The development of a wireless sensor network sensing node utilising adaptive self-diagnostics. In *Proceeedings of the Second International Workshop on Self-Organizing Systems*.

[Li et al., 2006] Li, M., Ganesan, D., and Shenoy, P. (2006). PRESTO: Feedback-driven data management in sensor networks. In *Proc. of the NSDI'06*.

[Lin and Gerla, 1997] Lin, C. and Gerla, M. (1997). Adaptive clustering for mobile wireless networks. *IEEE Journal of Selected areas in Communications*, 15(7).

[Lindsey et al., 2002] Lindsey, S., Raghavendra, C., and Sivalingam, K. M. (2002). Data gathering algorithms in sensor networks using energy metrics. *IEEE Transactions on Parallel and Distributed Systems*, 13(8):924–936.

[Lindsey et al., 2001] Lindsey, S., Raghavendra, C. S., and Sivallingam, K. (2001). Data gathering in sensor networks using the energy*delay metric. In *Proceedings of the IPDPS Workshop on Issues in Wireless Networks and Mobile Computing*.

[Lu et al., 2005] Lu, C. et al. (2005). A spatiotemporal query service for mobile users in sensor networks. In *Proceedings of the 25th International Conference on Distributed Computing Systems (ICDCS'05)*.

[Luo et al., 2005] Luo, H., Luo, J., and Liu, Y. (2005). Energy efficient routing with adaptive data fusion in sensor networks. In *Proceedings of the Third ACM/SIGMOBILEe Workshop on Foundations of Mobile Computing'05*.

[Madden et al., 2005] Madden, S. et al. (2005). Tinydb: An acquisitional query processing system for sensor networks. *ACM Transactions on Database Systems*, 30(1).

[Madden et al., 2002] Madden, S., Franklin, M. J., Hellerstein, J., and Hong, W. (2002). Tag: A tiny aggregation service for ad-hoc sensor network. In *Proc. of the Fifth USENIX Symposium on Operating Systems Design and Implementation*.

[Mainland et al., 2005] Mainland, G., Parkes, D., and Welsh, M. (2005). Decentralized, adaptive resource allocation for sensor networks. In *Proc. of the NSDI'05*.

[Manjeshwar and Agrawal, 2001] Manjeshwar, A. and Agrawal, D. P. (2001). TEEN: A protocol for enhanced efficiency in wireless sensor networks. In *Proceedings of the first International Workshop on Parallel and Distributed Computing Issues in Wireless Networks and Mobile Computing*.

[Manjeshwar and Agrawal, 2002] Manjeshwar, A. and Agrawal, D. P. (2002). Apteen: A hybrid protocol for efficient routing and comprehensive information retrieval in wireless sensor networks. In *Proceedings of the 2nd International Workshop on Parallel and Distributed Computing Issues in Wireless Networks and Mobile Computing*.

[Marbini and Sacks, 2003] Marbini, A. and Sacks, L. (2003). Adaptive sampling mechanisms in sensor networks. In *London Communications Symposium*.

[Martinez et al., 2004] Martinez, K., Hart, J. K., and r. Ong (2004). Environmental sensor networks. *IEEE Computer Magazine*, 37(8):50–56.

[Mhatre et al., 2004] Mhatre, V. et al. (2004). A minimum cost heterogeneous sensor network with a lifetime constraint. *IEEE Transaction on Mobile Computing*.

[Min and Chandrakasan, 2003] Min, R. and Chandrakasan, A. (2003). Mobicom poster: top five myths about the energy consumption of wireless communication. *Mobile Computing and Communications Review*, 7(1):65–67.

[Moore et al., 2004] Moore, D., Leonard, J., Rus, D., and Teller, S. (2004). Robust distributed network localization with noisy range measurements. In *Proc. of ACM SenSys 2004*.

[Muir and Garcia-Luna-Aceves, 1998] Muir, A. and Garcia-Luna-Aceves, J. J. (1998). An efficient packet sensing mac protocol for wireless networks. *Mobile Networks and Applications*, 3(2):221–234.

[Mukhopadhyay et al., 2004a] Mukhopadhyay, S., Panigrahi, D., and Dey, S. (2004a). Data aware, low cost error correction for wireless sensor networks. In *Proceeedings of 2004 IEEE Wireless Communications and Networking Conferenc*.

[Mukhopadhyay et al., 2004b] Mukhopadhyay, S., Panigrahi, D., and Dey, S. (2004b). Model based error correction for wireless sensor networks. In *Proceeedings of the 1st Annual IEEE Communications Society Conference on Sensor and Ad Hoc Communications and Networks*.

[Nath et al., 2007] Nath, S., Liu, J., and Zhao, F. (2007). Sensormap for wide-area sensor webs. *Computer*, 40(7):106–109.

[Peterson and Davie, 2003] Peterson, L. L. and Davie, B. (2003). *Computer Networks: A Systems Approach.* Morgan Kaufmann.

[Picconi et al., 2006] Picconi, F. et al. (2006). Probabilistic validation of aggregrated data for v2v traffic information systems. In *Proc. of the Third ACM International Workshop on Vehicular Ad Hoc Networks (VANET 2006).*

[Polastre et al., 2004] Polastre, J., Hill, J., and Culler, D. (2004). Veratile low power media access for wireless sensor networks. In *Proc. of ACM SenSys 2004.*

[Polastre et al., 2005] Polastre, J., Szewczyk, R., and Culler, D. (2005). Telos: Enabling ultra-low power wireless research. In *Proc. of the 14th International Conference on Information Processing in Sensor Networks: Special track on Platform Tools and Design Methods for Network Embedded Sensors (IPSN/SPOTS).*

[Pottie and Kaiser, 2000] Pottie, G. and Kaiser, W. (2000). Wireless integrated network sensors. *Communications of the ACM*, 43(5):51–58.

[Rajendran et al., 2003] Rajendran, V., Obraczka, K., and Garcia-Luna-Aceves, J. J. (2003). Energy-efficient, collision-free medium access control for wireless sensor networks. In *Proc. of ACM SenSys 2003.*

[Ramakrishnan, 1998] Ramakrishnan, R. (1998). *Database Management Systems.* WCB/McGraw-Hill.

[Sadagopan et al., 2003] Sadagopan, N. et al. (2003). The acquire mechanism for efficient querying in sensor networks. In *Proceedings of the First International Workshop on Sensor Network Protocol and Applications.*

[Sadagopan and Krishnamachari, 2004] Sadagopan, N. and Krishnamachari, B. (2004). Maximizing data extraction in energy-limited sensor networks. In *Proceedings of IEEE INFOCOM 2004*.

[Sadler et al., 2004] Sadler, C., Zhang, P., Martonosi, M., and Lyon, S. (2004). Hardware design experiences in zebranet. In *Proc. of ACM SenSys 2004*.

[Santi and Chessa, 2002] Santi, P. and Chessa, S. (2002). Crash faults identification in wireless sensor networks. *COMPUTER COMMUNICATIONS*, 25(14):1273–1282.

[Schurgers and Srivastava, 2001] Schurgers, C. and Srivastava, M. (2001). Energy efficient routing in wireless sensor networks. In *MILCOM Proceedings on Communications for Network-Centric Operations: Creating the Information Force*.

[Seada et al., 2004] Seada, K., Zuniga, M., Helmy, A., and Krishnamachari, B. (2004). Energy-efficient forwarding strategies for geographic routing in lossy wireless sensor networks. In *Proc. of ACM SenSys 2004*.

[SenseWeb,] SenseWeb. Microsoft research.

[Sha et al., 2006a] Sha, K., Du, J., and Shi, W. (2006a). Wear: A balanced, fault-tolerant, energy-efficient routing protocol for wireless sensor networks. *International Journal of Sensor Networks*, 1(2).

[Sha et al., 2008a] Sha, K. et al. (2008a). Data quality and failures characterization of sensing data in environmental applications. In *Proceedings of CollaborateCom 2008*.

[Sha et al., 2008b] Sha, K. et al. (2008b). Modeling data consistency in wireless sensor networks. In *Workshop Proceedings of ICDCS 2007 (WWASN 2007*.

[Sha et al., 2008c] Sha, K. et al. (2008c). Spa: Smart phone assisted chronic illness self-management with participatory sensing. In *Proceedings of The 2nd International Workshop on Systems and Networking Support for Healthcare and Assisted Living Environments (Healthnet'08), in conjunction with MobiSys08*.

[Sha and Shi, 2004] Sha, K. and Shi, W. (2004). Revisiting the lifetime of wireless sensor networks. In *Proceedings of ACM SenSys'04*.

[Sha and Shi, 2005] Sha, K. and Shi, W. (2005). Modeling the lifetime of wireless sensor networks. *Sensor Letters*, 3(2):126–135.

[Sha and Shi, 2006a] Sha, K. and Shi, W. (2006a). On the effects of consistency in data operations in wireless sensor networks. In *Proceedings of IEEE 12th International Conference on Parallel and Distributed Systems*.

[Sha and Shi, 2006b] Sha, K. and Shi, W. (2006b). Using wireless sensor networks for fire rescue applications: Requirements and challenges. In *Proceedings of IEEE EIT 2006*.

[Sha and Shi, 2008] Sha, K. and Shi, W. (2008). Consistency-driven data quality management in wireless sensor networks. *Journal of Parallel and Distributed Computing*, 68(9):1207–1221.

[Sha et al., 2006b] Sha, K., Shi, W., and Sellamuthu, S. (2006b). *Load Balanced Query Protocols for Wireless Sensor Networks*. Wiley-IEEE Press.

[Sha et al., 2004] Sha, K., Zhu, Z., and Shi, W. (2004). Capricorn: A large scale wireless sensor network simulator. Technical Report MIST-TR-2004-001, Wayne State University.

[Shah and Rabaey, 2002] Shah, R. and Rabaey, J. (2002). Energy aware routing for low energy ad hoc sensor networks. In *Proceedings of the IEEE Wireless Communications and Networking Conference (WCNC'02)*.

[Shah et al., 2003] Shah, S., Dharmarajan, S., and Ramamritham, K. (2003). An efficient and resilient approach to filtering and disseminating streaming data. In *Proc. of 29th International Conference on Very Large Data Bases.*

[Sharaf et al., 2004] Sharaf, M., Beaver, J., Labrinidis, A., and Chrysanthis:, P. (2004). Balancing energy efficiency and quality of aggregate data in sensor networks. *The VLDB Journal*, 13(4):384–403.

[Shekhar et al., 2003] Shekhar, S., Lu, C., and Zhang, P. (2003). A unified approach to spatial outliers detection. *Geoinformatica*, 7(2).

[Shih et al., 2001] Shih, E. et al. (2001). Physical layer driven protocol and algorithm design for energy-efficient wireless sensor networks. In *Proceedings of the 7th Annual ACM/IEEE International Conference on Mobile Computing and Networking (MobiCom'01)*, Rome, Italy.

[Shnayder et al., 2004] Shnayder, V. et al. (2004). Simulating the power consumption of large-scale sensor network applications. In *Proc. of ACM SenSys 2004.*

[Shrivastava et al., 2004] Shrivastava, N., Buragohain, C., Suri, S., and Agrawal, D. (2004). Medians and beyond: New aggregation techniques for sensor networks. In *Proc. of ACM SenSys 2004.*

[Shrivastava et al., 2006] Shrivastava, N. et al. (2006). Target tracking with binary proximity sensors: Fundamental limits, minimal descriptions, and algorithms. In *Proc. of ACM SenSys 2006.*

[Simon et al., 2004] Simon, G., Ledeczi, A., and Maroti, M. (2004). Sensor network-based countersniper system. In *Proc. of ACM SenSys 2004.*

[Skordylis et al., 2006] Skordylis, A., Guitton, A., and Trigoni, N. (2006). Correlation-based data dissemination in traffic monitoring sensor networks. In *Proceedings of the 2006 ACM CoNEXT conference*.

[SNA,] SNA. A network architecture for wireless sensor networks.

[Srivastava et al., 2006] Srivastava, M. et al. (2006). Wireless urban sensing systems. Technical Report CENS Technical Report No. 65, CENS.

[Ssu et al., 2006] Ssu, K. et al. (2006). Detection and diagnosis of data inconsistency failures in wireless sensor networks. *Computer Networks: The International Journal of Computer and Telecommunications Networking*, 50(9).

[Sutton and Barto, 1998] Sutton, R. and Barto, A. (1998). *Review of Reinforcement Learning: An Introduction*. The MIT Press.

[Szewczyk et al., 2004a] Szewczyk, R. et al. (2004a). Habitat monitoring with sensor networks. *Communications of the ACM*, 47(6):34–40.

[Szewczyk et al., 2004b] Szewczyk, R., Mainwaring, A., Polastre, J., and Culler, D. (2004b). An analysis of a large scale habitat monitoring application. In *Proc. of ACM SenSys 2004*.

[Tanenbaum and van Steen, 2002] Tanenbaum, A. S. and van Steen, M. (2002). *Distributed Systems: Principles and Paradigms*. Prentice-Hall.

[Tang and Xu, 2006] Tang, X. and Xu, J. (2006). Extending network lifetime for precision-constrained data aggregation in wireless sensor networks. In *Proc. of IEEE Conference on Computer Communications (INFOCOM'06)*.

[Tatbul et al., 2004] Tatbul, N. et al. (2004). Confidence-based data management for personal area sensor networks. In *Proceeedings of the 1st international workshop on Data management for sensor networks: in conjunction with VLDB 2004*.

[Thiemann et al., 2008] Thiemann, C., Treiber, M., and Kesting, A. (2008). Estimating acceleration and lane-changing dynamics based on ngsim trajectory data. In *Transportation Research Board Annual Meeting*.

[Traffic Simulator,] Traffic Simulator. Traffic simulator.

[Trappe et al., 2005] Trappe, W., Zhang, Y., and Nath, B. (2005). Miami: methods and infrastructure for the assurance of measurement information. In *Proceedings of the 2nd international workshop on Data management for sensor networks*.

[VII,] VII. Vehicle infrastructure integration.

[Wavescope,] Wavescope. Wavescope: An adaptive wireless sensor network system for high data-rate applications.

[Welsh and Mainland, 2004] Welsh, M. and Mainland, G. (2004). Programming sensor network using abstract regions. In *Proceedings of the First USENIX/ACM Networked System Design and Implementation*.

[Woo and Culler, 2001] Woo, A. and Culler, D. (2001). A transmission control scheme for media access in sensor networks. In *Proceedings of the 7th Annual ACM/IEEE International Conference on Mobile Computing and Networking (MobiCom'01)*, Rome, Italy.

[Xu et al., 2004] Xu, N. et al. (2004). A wireless sensor network for structural monitoring. In *Proc. of ACM SenSys 2004*.

[Xu et al., 2001a] Xu, Y., Heidemann, J., and Estrin, D. (2001a). Geography-informed energy conservation for ad hoc routing. In *Proceedings of the 7th Annual ACM/IEEE International Conference on Mobile Computing and Networking(MobiCom'01)*.

[Xu et al., 2001b] Xu, Y., Heidemann, J., and Estrin, D. (2001b). Geography-informed energy conservation for ad-hoc routing. In *Proceedings of the 7th Annual ACM/IEEE International Conference on Mobile Computing and Networking(MobiCom'01)*.

[Xue and Ganz, 2004] Xue, Q. and Ganz, A. (2004). On the lifetime of large scale sensor networks. *Elsevier Computer Communications*.

[Yao and Gehrke, 2002] Yao, Y. and Gehrke, J. (2002). The cougar approach to in-network query processing in sensor networks. *ACM SIGMOD Record*, 31(3):9–18.

[Ye et al., 2004] Ye, F. et al. (2004). Statistical en-route filtering of injected false data in sensor networks. In *Proceeedings of IEEE Infocom 2004*.

[Ye et al., 2005] Ye, F. et al. (2005). Statistical en-route filtering of injected false data in sensor networks. *IEEE Journal on Selected Areas in Communications*, 23:839–850.

[Ye et al., 2002] Ye, W., Heidemann, J., and Estrin, D. (2002). An energy-efficient mac protocol for wireless sensor networks. In *Proc. of IEEE Conference on Computer Communications (INFOCOM'02)*, New York, NY.

[Younis et al., 2002] Younis, M., Youssef, M., and Arisha, K. (2002). Energy-aware routing in cluster-based sensor network. In *Proceedings of ACM/IEEE MASCOTS'2002*.

[Younis and Fahmy, 2004] Younis, Q. and Fahmy, S. (2004). Distributed clustering in ad-hoc sensor networks: A hybrid, energy-efficient approach. In *Proc. of IEEE Conference on Computer Communications (INFOCOM'04)*.

[Yu and Vahdat, 2000] Yu, H. and Vahdat, A. (2000). Design and evaluation of a continuous consistency model for replicated services. In *Proceedings of the OSDI'2000*.

[Yu et al., 2002] Yu, Y., Estrin, D., and Govindan, R. (2002). Geographical and energy-aware routing: A recursive data dissemination protocol for wireless sensor networks. Technical Report Computer Science Department Technical Report UCLA/CSD-TR-01-0023, UCLA.

[Zhang and Hou, 2004] Zhang, H. and Hou, J. (2004). On deriving the upper bound of α-lifetime for large sensor networks. In *Proceedings of the Fifth ACM International Symposium on Mobile Ad Hoc Networking and Computing*.

[Zhang and Cao, 2004] Zhang, W. and Cao, G. (2004). Optimizing tree reconfiguration for mobile target tracking in sensor networks. In *Proceedings of INFOCOM'04*.

[Zhuang et al., 2007] Zhuang, Y. et al. (2007). A weighted moving average-based approach for cleaning sensor data. In *Proceedings of the 27th International Conference on Distributed Computing Systems*.

ABSTRACT

ORCHIS: CONSISTENCY-DRIVEN ENERGY EFFICIENT DATA QUALITY MANAGEMENT IN SENSING SYSTEMS

by

KEWEI SHA

DECEMBER 2008

Advisor: Dr. Weisong Shi

Major: Computer Science

Degree: Doctor of Philosophy

As new fabrication and integration technologies reduce the cost and size of wireless sensors, the observation and control of our physical world will expand dramatically using the temporally and spatially dense monitoring afforded by wireless sensing systems. Their success is nonetheless determined by whether the sensor networks can provide a *high quality stream of data* over a long period. However, most previous efforts focus on devising techniques to save the sensor node energy and thus extend the lifetime of the whole sensor network. With more and more deployments of real sensor systems, in which the main function is to collect interesting data and to share with peers, data quality has been becoming a more important issue in the design of sensor systems. In this dissertation, we envision that the quality of data should reflect the timeliness and accuracy of collected data that are presented to interested recipients who make the final decision based on these data. Thus, we undertake a novel approach that detects deceptive data through considering the consistency requirements of data, and study the relationship between the quality of data and the multi-hop communication and energy-efficient design of networked sensor systems.

In this dissertation, we tackle the data quality management problem by proposing a general framework, called Orchis, which mainly consists of six components, including an analysis to the characteristics of the sensing data from an environmental application, a set of data consistency models customized to wireless sensing systems, a set of APIs to management the quality of collected data, an adaptive protocol for data sampling, a framework to detect and filter deceptive data, and a formal model for the lifetime of the wireless sensing system to evaluate the energy efficiency performance of the protocols. The experiment performance from both simulation and prototype shows that the Orchis framework is promising in terms of both energy efficiency and data consistency.

AUTOBIOGRAPHICAL STATEMENT

Kewei Sha is a Visiting Assistant Professor at Oklahoma City University. He got B.S. degree, major in Computer Science, from East China University of Science and Technology, 2001, and he got M.S. degree, major in Computer Science, from Wayne State University, 2006. His research interests focus on Distributed Systems, Wireless Sensor Networks, Vehicular Networks, Participatory Sensing, and Mobile Computing. His research topics includes data quality management, security and privacy, system protocols, and system design and modeling.